The Last Farm

A Novel of a Farm Family in the Era of Economic Development

J.A. Stein

J.A. Stein Publishing

Copyright © 2024 by J.A. Stein

All rights reserved.

No part of this publication may be reproduced, distributed, or transmitted in any form or by any means, including photocopying, recording, or other electronic or mechanical methods, without the prior written permission of the publisher, except as permitted by U.S. copyright law. For permission requests, contact J.A. Stein at authorjastein@gmail.com.

The story, all names, characters, and incidents portrayed in this production are fictitious. No identification with actual persons (living or deceased), places, buildings, and products is intended or should be inferred.

Print ISBN: 979-8-9864908-4-7

Book Cover by Venom Co.

Edited by Gail Delaney

Chapter 1

Bill Fogel stared at the sheaf of papers in front of him as if it would bite. Sweat beaded on his weathered, sun-tanned face. The lines of age merged with the stubble of a grey five o'clock shadow. He rubbed a hand over the bristle absently, only to startle at the unfamiliar sensation. How long had he been at this? He glanced at the clock and groaned, rising from his chair to look out the window over the rolling fields of corn as they burned golden in a late August sunset. The sun lit up the rolling hills in a green-blue haze, the bright pinks of clouds streaking overhead.

The beauty only made the decision harder.

Bill looked over his shoulder at the papers on the table, then abruptly strode to them and closed the folder of legal jargon that was somehow still not thick enough to encompass the enormity of what they were asking of him.

Years of hard work and history rested in this soil, three hundred and sixty-seven acres that his ancestors had cleared by hand of trees and rocks. And still the trees encroached into the fence lines, shading good ground and stunting crops. Still the rocks seemed to dig themselves from the belly of the earth, requiring the labor of younger men than he, as tractor bucket after tractor bucket of stones expanded the piles that marked the boundary of the farm. That is, if he could find young men willing to do the sort of labor they didn't even ask prisoners to do anymore.

Bill set the pen on the folder in a vertical line, unused. He took a quick drink of water from the kitchen sink, then pulled on his worn ball hat and pushed out the screen door, one that his dad had made back in the fifties. It made a solid smack-smack noise as it slapped shut behind him, but Bill was already off the porch and striding up the rutted dirt road that led to the overlook. As if the view from the kitchen window wasn't good enough, he needed to see the whole farm, the three-sixty view on the little hill where he'd once gotten down on one knee to propose.

That seemed a lifetime ago. It was...and yet he still remembered that day perfectly. He'd sweated all day, wondering if she'd say yes. He didn't remember the sunset from that day, but he did remember the way it brought out the auburn in her hair. He also remembered the way the light shown on the silver strands of her later years, haloing her as if she were an angel. He swallowed. A lifetime of sunsets had passed, as had his beautiful wife.

He strode up the hill faster, hands deep in pockets, ignoring the pain in his arthritic knees.

"Pap! Pap!" a woman's voice called from the house.

Bill instantly turned around, his eyes lighting on his twenty-five-year-old granddaughter, grown every bit as pretty as her mother and her grandmother before her. She jogged up the hill in a tank top and blue jeans, ponytail swinging in rhythm with her footsteps. She was yet another reminder of time's passage—how had his giggling granddaughter turned into the solemn-faced woman before him?

"Hi, honey. Everything okay?" He staggered backwards a step as she flung her arms around him in a hug.

"Mom just told me." Evie's voice was a mere whisper, shaky with emotion. "Are you really gonna sell?"

Bill blinked back his own emotion and hugged her back. "It's probably the right decision. For the whole family." He sighed, pushing her gently away that he could look her in the eye. "Ten million dollars is a lot of money."

He watched Evie's eyes go wide.

"Is that how much?" she asked.

He chuckled. "Your mother didn't tell you that?"

She shook her head.

"Well, there you have it." He took off his hat and scratched his head. "I'm walking up to the overlook. Want to come?"

Evie followed silently, her hands in her pockets. At the top they settled on the old wooden bench. Bill noted how it took a little longer than it used to for him to catch his breath.

"So, are you gonna take the money?" Sell was the unspoken word. She spoke with the same sympathetic tone one used when inquiring if one was going to euthanize a dog.

"Haven't decided yet." They watched the sunset in all its glory. Three hundred sixty-seven acres of prime farmland lay before them. Hard to imagine what it would look like when it was bulldozed for a warehouse. Would this little rise of hilltop even remain? When they offered well above the market value to a farmer who busted his back every year for less than minimum wage—if the weather cooperated—it was mighty tempting. "What d'you think I should do?"

Evie sighed, her face tight with thought. "Ten million, eh? And how long has this place been in the family?"

"Two hundred years. I think. Maybe longer, if the story that we bought it from the Indians is true."

"We bought it from the Indians?"

"My great-great—I don't know how far back really—but he came over that mountain there in 1748." Pap pointed to the long ridge the settlers had nicknamed the Blue Mountain centuries before. Even today, it had a bit of the blue haze the settlers had named it for. The name had stuck, as did the mountain's wildness. To this day there were only a few roads crossing it, and the Appalachian Trail ran along its crest. "So they say. This land wasn't part of the treaty yet in those days. That was before the French and Indian War, you see. When my dad applied for this place to be given centennial farm distinction years back, he did all the research and came up with a deed from around 1769. But family legend holds that the original ancestor was a pioneer who bought the land straight from the Indians themselves, before the governor managed to finalize the treaties."

"Really? I never heard this story."

Bill chuckled. "There's a lot of stories about this place you probably haven't heard. You didn't run wild here like I did when I was a kid. Like your mom and aunt did."

Evie bit her lip. "Too busy with soccer and swimming and dance class, huh?"

"Don't get me wrong, it's wonderful that you got to do all those things." Bill put his arm around her and pulled her close. "We worked hard that your generation could have those opportunities, because when I was growing up, they just didn't exist. Not 'round here at least."

Evie rested her head on his shoulder. "I guess I better get caught up then—on all those stories. Before it's too late."

"Want to come back in the morning? I'll give you the grand tour."

"That sounds great, Pap."

Bill's head spun as the tales he'd heard over the years bubbled to the surface. The places he wanted to show Evie flashed like a slideshow behind his eyes. Were they even still there? Or had nature reclaimed them? And there was one story he would have to look up tonight. It had been so long, he had to check the facts. Didn't want Evie to feel misled. But if his memory was still right, it was a tale she was going to love.

He gave her shoulder a squeeze, watching as a redtail hawk cut through the sky, its belly yellow with golden light. Its screech cut the air, as wild as the forests on the mountain behind him, as free as the wind that caressed the corn stalks before them with a whisper. His decision could wait another day. For now, he had only to live in the present.

Chapter 2

Evie spent the next morning wandering along the creek with Pap, where he showed her the springhouse with its partially-collapsed roof. Inside, a nice fat water snake made her jump about five feet and shriek. Pap had laughed at her. Then there was the supposed swimming hole, which seemed far too scummy to even dip a toe into, much less swim in, particularly if that water snake liked to travel any distance. In the bottom of the old bank barn, the dairy stalls sat empty as they had for years, dust and dead rodents creating a whole new layer of earth inside the building. The names of the cows that had once occupied each stall still hung on dusty, faded cardboard signs on the beam above. Evie had seen all this before, but the finality and forcefulness with which Pap now led his tour sharpened her attention.

Should she be recording this tale?

As he talked, she opened a cabinet, bracing for a mouse to run out, but was instead only greeted by the sharp smell of turpentine and other thirty-year-plus old animal medications. The whole farm was like a time capsule. Certain places had been undisturbed for decades. The value of this barn's holdings to an antique dealer was not lost to her. Pap was still pointing out things as he led her back out of the barn. There was the big hay hook they used before they bought the hay elevator. She could only imagine the effort involved with hefting each individual bale into the hay mound with human power. There was the patched floorboard where Great-Uncle Tommy had fallen through and cut his leg. Those pigeons were nearly impossible to get rid of. They'd been trying for forty years.

Evie listened, hearing the words but more importantly hearing her grandfather's need to relay all this history before it was bulldozed. With him gone, the buildings gone, these oral stories would be forgotten. She wished she'd had the foresight to bring a recorder, or at least a pen and paper, along on this walk. She'd been a frequent visitor to the farm her whole life. She had her own stories tied to places. There was the spot under the oak tree where she'd sat and collected acorns and there, her and her cousins' hut in the old chicken coop, an old curtain still hanging with the cobwebs. She could fill a book with her own memories, and Pap's went back far deeper, far longer. He was also the last caretaker of stories from the farmers before him. That kind of depth of memory was overwhelming to comprehend.

It was going to be hard for him to let all this go.

Pap leaned on the fencepost in the barnyard, struggling to catch his breath, sweat glistening on his forehead in the early afternoon sun, which was peaking at a scorching eighty-five degrees—a last warm day of summer as September settled in. His age was catching up with him, though he was far spryer than most seventy-seven-year-olds. At times Evie wondered if she should worry about him, being alone out on the farm like this. Then she watched him hop on a tractor and putt away like a man twenty years younger.

Today, he seemed his age.

"Let's take a break. Get lunch," Evie suggested.

He nodded and followed her up to the house. Evie set about pulling the sandwich ingredients out of the fridge and onto the table. She filled two tall glasses with instant iced tea from the pitcher. Pap downed his in one gulp, then refilled his glass.

"Sweetie, go grab that book on the counter there." He pointed to a red book with darkened edges.

Evie wiped her hands on her jeans, a bite of her haphazardly compiled sandwich already in her mouth. Pap was making his sandwich in a much neater fashion, watching her with a mischievous glimmer in his eye. She glanced at the book and back to him.

"What are you up to?"

He nodded at the book. "The page is bookmarked. Careful—that book is over a hundred years old now."

"*Old Schuylkill Tales,*" Evie read aloud. She shoved another huge bite of sandwich into her mouth, then wiped her hands before carefully cracking the pages at the bookmark.

"Go on. Read it."

Evie scanned the paragraph about an Indian hiding gold from his friends and then being murdered for it, his ghost wandering the banks of the Swatara Creek searching for the treasure. For her grandfather's sake, she didn't roll her eyes. But the next paragraph made her hesitate. She read, then reread it. "'Gold was said to have been found upon the Gobbleberg, and the Indian superstition claimed that when it thundered and lightninged the rocks were sometimes cleft in twain and the hidden recesses were discovered to be gorged with nuggets of gold.' What is this?" She scanned the text again, a memory from her childhood vividly in her mind. The lightning bolt had crashed down right next to her, blinding her, knocking her off her feet. She'd felt the static lift every hair on her body, the heat of the earth filling her nose as she'd staggered blindly back down the hill to the farmhouse, her panicked mother scooping her up into the dry house where her tears were soothed as her sight returned. She could still feel that heat—and that fear—every time it thundered.

Pap watched her carefully, chewing his sandwich. "You and I both know that lightning hits the mountain on the backside of the farm at a much greater frequency than anywhere else. I never found any gold, but it makes for an interesting story, doesn't it?"

Evie snapped the book shut and rolled her eyes freely now. "Gold? In Schuylkill County? Nothin' here but black gold, and that's to the north. Which reminds me, I need to order more coal for the shop before the weather turns cold."

Pap shrugged, turning back to his sandwich, washing it down with the tea. Evie had no interest in hers at the moment.

"You really think—"

"I don't think anything, but it's a good story."

"You never looked?"

"I was runnin' a farm!"

Evie hesitated. "I haven't been back up there since I was a kid."

"I know."

"That day really scared me."

"I know."

"But you think..."

"I'm too old to climb up there myself. That's a good hike. And the four-wheeler path cuts off a good ways from the top." He grumbled to himself, "Darn tree still needs to be cut..."

"But it would be good to know, you know, before the farm is sold, if there are gold nuggets on that hill."

"You could probably convince your cousin Jethro to lend you his metal detector."

"Ha!" She snorted, the tension ebbing. She plopped down in a chair and filled her mouth with sandwich. "Anywhere that detector goes, Jeth goes. And he's annoying." She mumbled to herself, her mouth full, "Fancy schmancy engineer..."

"You've only got two cousins, sweetheart. You better figure out how to get along with both of them." Pap shrugged, a smile on his lips. At that moment the phone rang, and he retrieved it from where it hung on the wall down the hall.

Evie leaned her chair back to watch him, her sandwich steadily disappearing.

"I'm going to need a few more weeks to make the decision." Pause. "Well, you've waited two hundred years for this property to come up for sale. What's two weeks?" Pause. Pap snorted, his tone impatient now. "If the property is so valuable to you, then maybe I should entertain more offers. From your competitors!" He hung up the phone.

Evie watched wide-eyed as he returned. "They getting impatient?"

"You could say that. I mean, the paperwork's only been on my desk a month. I don't see what the hurry is."

"A month is a long time to those people." Evie knew. She was running her own small business in town, a bookstore. She couldn't imagine the frustration of having no response for a whole month on a business decision. Particularly one of this scale. Then again, these things took time. She wasn't making any ten-million-dollar decisions.

Pap took off his hat and rubbed the top of his grey head. "Look, Evie. Call your cousin. Go for a hike. See what you find up there. You're a big kid now. You can hike up the mountain. Who knows, maybe it will be the last time you get the chance. Just be careful, okay?"

"Yes, Pap." She did roll her eyes this time. "I'll try to be brave and not let the lightning eat me. Just to look for mythical golden nuggets."

He gave her shoulder a squeeze. "Love ya, sweetie."

She sobered a bit. "Love you, too, Pap."

Chapter 3

Jeth's long strides had Evie's legs aching in her attempt to keep up before they even reached the end of the field. "Can you spare me a moment?" she panted. "This isn't a freakin' marathon."

Jeth laughed. "City legs!"

"I have an office job."

"Like I said." Jeth shrugged, but he did at least pause to face her before he started walking backwards up the gradual slope. His metal detector was casually slung over a shoulder, his arm draped over it.

"And how exactly is it you stay so fit? You have an office job, too," Evie whined. Blast him, they'd grown up more as siblings than cousins, and the rivalry still ran strong.

"And a gym membership, cuz." His bright smile that had charmed all the girls in their high school only made her want to claw his eyes out. When he flexed his biceps in jest she picked up a rock and threw it at him. It missed, of course.

Jeth chuckled and ran up the last of the dirt road to the trees that edged the field. Behind him, an overgrown path wound up the steep hillside into the woods. Evie took a swig of water and grumbled to herself. She really was out of shape. She'd never been fat per se, but hours behind a desk with coffee and chips for lunch and dinner were starting to take their toll on her poor muscles.

"Congratulations, you've made it to the tree line. A quarter-mile down, now the big-ole climb to go." Tree line, ha. They were far from any elevation that involved true tree line, but they *were* at the edge of the field. Jeth stepped into the forest.

"Why did I invite you?" Evie asked herself.

Jeth replied anyway. "Because I've got the metal detector!" His voice carried into the woods as he raised it above his head like a trophy. They picked over the rocky path, carefully watching their footsteps to avoid twisting ankles.

Evie sealed her lips into a thin line, forcing herself to match her cousin's pace. It was only a twenty-minute hike to the top. She could do it.

The fact that she was making the climb for the first time in over a decade wasn't lost on her. She thought about the whipping winds of the storm, the flash of lightning, the crack of thunder in her ears. It was enough to make her stop and close her eyes. Now it was her heartbeat that thudded in her ears.

"Poor Evie. So out of shape," Jeth taunted from up ahead.

She didn't dare admit her true fears, lest he really have something to make fun of her for. She plunged ahead with gritted teeth. *Be an adult,* she told herself.

Sweating from the humid summer heat and with her legs burning, Evie emerged on top of the knoll. Contrary to what one might expect from the highest point on the little farm, there was absolutely no view. The trees grew thick here, the steep slope and huge boulders too rugged to make it worth timbering, unlike the rest of the farm. She sat on a rock and took another drink. Unlike the last time she'd been up on the hill, the air was quiet and the sun was shining. No hint of storms.

No hint of gold either.

"All I see is granite." She looked around at a small river of boulders, trees springing from in and around them like the bristle of a porcupine's back. Thick moss and lichen covered most. Beyond the outcropping of rock, mountain laurel and stubby hemlocks grew in patches, leaving narrow paths like the one they'd followed weaving over the leafy earth.

Jeth stood on a boulder, hands on his hips like a superhero, though his brow furrowed as he studied the rock. "I've been up here a hundred times, and I never saw gold. Though I wasn't looking for gold. I was looking for deer. Rabbits. Turkeys."

Evie picked at the gravel of the rock, then knocked at the boulder with a smaller rock. Neither rock chipped.

"You know those gold mine reality shows? How they're always chasing the quartz veins?" Jeth squatted down next to a boulder, studying it.

"Yeah." Evie frowned.

"Look closely. All of this granite is laced with quartz. It's just not in veins; it's in little pieces."

"It's always like that around here, though. Didn't you ever notice how much quartz is in Pap's creek?"

"Why would gold differentiate if it runs with quartz out west or here?"

Evie looked closer at the rock she sat on. The quartz shimmered back at her in the light. "I read once that gold nuggets were found in the Susquehanna River."

"That's a long way from here."

She shrugged. "But where did that gold come from?"

"Wouldn't your old boyfriend know?"

Evie straightened and glared at Jeth, her heart pounding at the mere mention of her ex. For once he avoided her eye and didn't comment further. He turned on the metal detector and poked around for a bit. Evie followed his lead in avoiding the subject, and dug into the earth with the trowel he provided, her pulse rising with each beep they found, only to fall flat when they uncovered old soda tabs or bottle caps. Eventually Evie became so frustrated she stopped digging, instead stretching out on a flat rock, staring up at the trees and their green leaves. The warmth of the rock on her back lulled her towards sleep, and she reconsidered her childish decision to avoid this place all these years.

Jeth continued beeping through the woods. Occasionally sound would cease as he dug in the leaves, only to hastily resume, sometimes after a frustrated expletive. After an hour he gave up and sat on the rock with Evie.

"What do you think about Pap selling the farm?" His tone of jest was gone, replaced by sadness.

"It breaks my heart. But our parents are getting close to retirement. They were never interested in farming anyway. You and me and your sister all have college degrees and *really* well-paying jobs. Well, the future *Doctor* Marissa *will* have a well-paying job. Hard to warrant trading that in for farm work. You know how hard Pap struggled all these years."

"I wish I could buy it." Jeth looked down at the rocks, picking at the loose gravel. "I'd buy you all out if I could. But I don't have the market value, let alone the industrial market value. And Pap's gonna need that money into his retirement now."

Evie shaded her eyes, studying him. He seemed sincere for once. But it was hard to picture him as a farmer. "Would you farm it?"

"Maybe lease it to another farmer until I can pay back the loan. My current job would pay it off way faster." He snorted. "Ironic, isn't it? The wannabe farmers can't afford land, and those of us that could afford it don't know how to farm."

"Does Pap know you'd be interested?"

"No. No point in getting his hopes up. I can't pay ten mill."

"Yeah. Anyway."

They watched as a squirrel rustled through the leaves in search of a nut.

"My mom says it's a wonderful thing. That Pap deserves a good retirement," Evie said quietly.

"I'm pretty sure my mom feels the same. Not that the farm matters to her. She's been out of state for years already."

"She and Uncle Rick love Florida. Nothing wrong with that."

"Maybe if my dad had been more interested in the farm, then he could have helped Pap out."

"Jeth, your dad worked his own kind of overtime. He deserves his quasi-retirement as much as Pap, if not more, after his years in corporate. How many birthday parties did he miss? Baseball games?"

Jeth threw down a stick that he'd been toying with. "I guess I'm turning out just like him, selling my life to my work." He abruptly stood and offered Evie a hand.

"Come on. While we're up here, let's go metal detect where we can find something."

Evie's brow furrowed in curiosity, and she followed. He led her partway down the backside of the hill, completely off trail now.

"This still Pap's land?" she questioned. Knee-high huckleberry bushes and ferns brushed her bare legs. She waved a hand at an errant fly.

"Yup. Down to the posted signs about halfway down the hill." Jeth stepped over a huge log, then pointed into the woods. Evie's eyes went wide at the tumbling stones of what once was a structure. The white posters could be seen another fifty yards beyond it.

"What is this?"

"Found this hunting a few years back. Always wanted to get the detector up here on it, see what I can find." As he switched on the metal detector, it immediately starting pinging. Evie poked around the ruin. The stones of the foundation and a crumbling chimney were all that was left of a tiny cottage. The roof was long gone. A tree that she could barely get her arms around grew up the center of the ruins.

"This has to be from the pioneer days." She looked at the ground and noted a large piece of metal. When she nudged it with her toe, an old plow blade appeared. A few steps away, a dry creek bed cut down the side of the mountain, its banks filled with leaves and fallen trees. "They must have had a spring that ran dry."

"I think fire made them move." Jeth threw a chunk of wood the size of his fist to Evie. She caught it. "What's left of the beams looks burned."

"I can't believe there's anything left of the beams." She eyed the wood, scrutinizing the solid texture. "This might not be from the cabin. Somebody just had a campfire here."

Jeth didn't answer as he focused in on a ping, digging into the loamy earth. He let out a sigh of frustration, throwing the find down. "You're probably right. Party spot. Just a bottlecap."

Evie went over and picked it up. The rusted cap showed the logo of a local brand that had gone out of business decades before. She snorted. "Yeah, 1970's party spot. I wonder if our moms brought their friends up here."

Another ping already distracted Jeth, this one low and steady. He knelt in the earth, pulling at roots with his trowel.

"Huh. That's a little more interesting." He handed the find to Evie. It was an old nail, unlike any she'd ever seen. It looked to be handmade. The sides were square, not round, and it was heavily rusted.

Jeth continued, catching another hit with a high-pitched ping. "Probably another bottle-cap," he mumbled. But even as Jeth dug deeper and the ping grew clearer, he still couldn't find the item. About a foot down into the soil, he pulled out a small, round object.

"Naw, it can't be…" he whispered. Then, "It is! Evie, check this out." He wiped more dirt away. "Now this gets more fun—"

"What?" she leaned over his shoulder.

"It's a coin. I can't see the date, but it's a coin. A really old one." He waved it under the detector. "Silver."

"Sheesh, they haven't made those in a while."

"Like, for about a hundred years." He skimmed the ground looking for more, but there were no more pings on the detector. "Well, it ain't gold, but it's somethin'."

"Hey, Jeth, what's that?" Evie stooped to pick up an oddly shaped rock from the hole he'd dug. As she brushed the dirt off, her pulse quickened. There in her hand was a perfect arrowhead, as straight and sharp as any she'd seen in a museum. "Wow."

"Oh, now that is cool." It was Jeth's turn to look over *her* shoulder. "Crazy how they were in the same hole."

Evie looked up to Jeth, excitement bubbling through her. She had arrowheads purchased from gift shops in a closet at home, but to find one on Pap's land out of a random hole was far more interesting. She pressed it into her palm. "Let's go show Pap what we found."

Chapter 4

When Evie and Jeth burst into Bill's house, he almost forgot that his two grandchildren were in their twenties. Their excitement was palpable, and their faces were as bright as children on Christmas morning. They weren't quite jumping up and down, but their cries of "Look, look!" held the same enthusiasm.

He felt a twinge of that youthfulness as he leaned over the sink to look at the finds they were rinsing under the faucet.

"Look, Pap. I found an old coin." Jeth held up the oxidized round object. Maybe it was a coin. Maybe it was a washer from a tractor bolt.

"Oh, that's nice." Bill smiled, patting Jeth on the back.

"Look what I found!" Evie elbowed Jeth out of the way. She held an arrowhead in the palm of her hand.

Bill narrowed his eyes at it, plucking it from her palm. "Now that's something."

"Hey, I dug the hole she found it in," Jeth reminded them.

"Good job. Both of you." Kids at Christmas. Complete with the bickering over their toys. Still, Pap couldn't tear his eyes from the arrowhead. It looked authentic. If they'd found it in the woods, it had to be. That was incredible. "You know, I haven't found one of these here since I was a kid."

"You found one, too?" Evie asked.

"Yup. It was in the creek. Up above the swimming hole, actually." Bill smiled to himself. He could remember that day like it was yesterday, not over sixty years ago. The summer sun had been bright. He and Tommy had been swimming, and he'd wandered to look for crayfish. He'd found a lot of them, even dropped one down his aunt's dress, but when the arrowhead appeared it was like he'd found treasure. "I looked for years after that, trying to find more, and never could. Where did you find this one?" He looked up to Jeth.

"Backside of the mountain, by the old cabin."

"That place still there?"

"Not much left. Found a nail there, too, but I left that up there. Along with a handful of bottlecaps.

Bill turned the arrowhead over in his palm, then handed it back to Evie. He reached for Jeth's find and squinted at it. On impulse, he pulled a box of baking soda from the cupboard and a glass. He sprinkled a little of the baking soda, filled it an inch with water, then dropped in the coin, swirling it in a circle until the worst of the dirt flaked away and the tarnished silver beneath shown through. Then he retrieved it, all three of them leaning over his hand.

"I think that's a date," Evie said.

"1748? 1759? It's definitely the seventeen hundreds." Jeth's tone of excitement was unmistakable.

Bill frowned. That was far older than his ancestors were known to have occupied the land. Were the ruins on the hill that old? Perhaps the occupants just had an old coin. But an arrowhead with it? That all indicated the Indian days. Was it a coincidence?

Bill took a sharp breath. He handed the coin to Jeth and walked out of the room as fast as his old joints would allow. In the living room, he bent over an old cabinet. He heard Evie and Jeth follow him in, watching. He again felt the thrill of Christmas morning when he located the old Bible. Straightening, he placed it on a side table and opened its tattered cover. The book was huge. He couldn't translate roman numerals well enough to say what century it was from, but he knew they wrote dates in regular numerals in the eighteen hundreds. It was rumored to have been brought over from the old country. He turned through the pages of printed German script to the very back pages, where in scrawled handwriting he could read the family tree his something-great-grandmother had inscribed.

The earliest date was 1830. He let his shoulders sag in disappointment.

Evie now leaned over his shoulder. "That is amazing." She ran a finger across the old script. "This is our whole family tree."

Pap shook his head. "Only part of it. It doesn't go back far enough." Nor, quite frankly, did it go ahead enough. The most recent three generations were written on clean white computer paper in a filing cabinet somewhere.

"What are you looking for?"

"Our oldest American-born ancestor." He sighed. "There were rumors we were living on this land before William Penn was given Pennsylvania, but I don't know how you'd confirm that." Bill pulled the book away and closed it with care. Evie's arrowhead still had his mind wandering the woods like a teenage explorer. Really, it didn't matter who had once lived on the farm. It would all be gone soon anyway. They would cut into the mountain for the shale to level the fields to build a pad for a five-hundred-thousand square foot building.

He had to show them his own secret treasure spot, where he'd found the arrowhead years ago, before that was gone as well. "Come on, you two. I want to show you where the Indians used to sit by the creek."

Evie and Jeth followed him out to the creek, still chattering between themselves and theorizing on what they had found up at the old homestead. Bill had to smile to himself, grateful for the company. With the girls grown and his wife Grace gone, the farm had seemed awful quiet as of late. Having the kids back was nice. Though at the moment, as his two grown grandchildren threatened to push each other into the scummy swimming hole, he also knew he'd be grateful for the solitude when they left again.

Evie shrieked as Jeth grabbed her around the waist and swung her towards the water. He placed her back on her feet at the last minute, laughing with glee. Evie's face was red with fury, and she starting swatting him any place she could reach.

"That." —Swat. — "Is." —Swat. — "Not Funny." —Swat. — "Jethro Fogel-Mueller!"

Jeth ran around the little pond, leaping across the narrow creek at the head of the little dam with ease. Evie started to follow him and then turned back to Bill.

"Pap, did you see what he did?" Evie pointed at her cousin like she was ten again.

Bill laughed, removing his hat to scratch the top of his head. "Honestly Evie, I'm surprised he hasn't thrown you in there well before now. My brother used to toss me in on a weekly basis every summer."

"Pap!" Evie protested, her hands on her hips, giving a remarkable imitation of her mother.

Jeth hollered from across the pond, "Does that mean I can throw her in? Happy to keep up with tradition and all that."

Bill shook his head. "You missed your chance, kid. You're supposed to act like grownups now."

Jeth stomped his feet in mock reflection of his younger self, then laughed as he straightened and ran back around the pond to Pap's side. He put his hands in the pockets of his jeans, mimicking Bill. Bill shook his head a little, yet the corner of his mouth still turned up.

"The spot's right there," Bill said and pointed with a finger.

Evie got there first and pulled off her sneakers to wade into the cold stream, wincing as she did so before exhaling a sigh of relief.

"Feels good after a hike, doesn't it?" Bill asked.

She nodded.

That was the very reason his ancestors had put the swimming hole where it was. He imagined it had run much clearer than it did now, before the silt had accumulated and blocked the outflow. Yet another project that no one had gotten around to in decades. He forced his eyes back to the creek. Evie and Jeth were poking around with curiosity but not much direction.

"It was right up here," Bill said, and bent to peer at a bend in the narrow creek bed. "I looked through my whole childhood, and most of your mothers', and I never found another. I doubt there's any left. But you two finding one on the mountain today...that means there's still history here."

Chapter 5

Histtory stayed on Evie's mind all that night. She tossed, turned, searched the Internet, counted sheep, and still managed to beat the sun out of bed. After a quick stop to check on the bookshop and drop off a list of opening instructions for her assistant, she headed to the library for its nine o'clock opening.

Twenty minutes. She could spare twenty minutes of research and still make it back to the shop early enough for the mid-morning rush.

The minutes ticked by as Evie traced the spine of each book on the library shelf. Once. Twice. A third time. There were three books already in her arms but still, she shook her head with dismay that there were not more. She looked around the library's expanse, the massive stacks of the reference section surrounding her. How could such a volume of information be reduced down to one shelf, and from that shelf only a few books relating to the 1700s in the area? She clutched the books to her chest, pulled her purse strap higher on her shoulder, and went off in search of a librarian. There had to be a section she was missing.

There wasn't.

The librarian pulled up the catalog, scanned it, then walked her over to the same section she'd come from. The quantity of books regarding Schuylkill County's early history was small, and books regarding the Indian times before that even sparser. The core history book of the area was *Old Schuylkill Tales*, which not only was largely comprised of word-of-mouth tales but had already been committed to Evie's memory by a late-night read.

"You don't have anything else? Maybe something in a restricted section that's older?" she asked the librarian.

The librarian looked down her nose over her bifocals. "You realize restricted sections only exist in Harry Potter, right?"

Evie was unamused. She glanced at her watch. She was already late.

The librarian sighed. "Sorry, this is it. We can look in the inter-library loan system. Or we have newspapers on microfilm. And our coal mining section is one of the best—"

"Wrong century," Evie mumbled, her eyes roaming the shelves a last time. "Well, can I take these out?" She held out the three small volumes, a history of the Iroquois, *The Blue Book of Schuylkill County* which was a collection of biographies of early settlers, and a generalized book about William Penn that Evie doubted would be helpful.

"Sorry, this is the reference section. You can't take these out, but I'm happy to make photocopies for you."

Of course, she should have known that. Evie looked at the old Blue Book and its weathered spine. The poor thing had been published in the early 1900s. The bibliophile in her winced. She couldn't put the old book through that. "No, I'll just take some notes. You have some paper I can borrow?" Next time she'd have to visit the library more prepared. She again glanced at the clock on her phone. She hoped her assistant had managed the opening alright. Surely she could handle things for a few more minutes. There were no missed calls, so that was a good thing.

Evie power-read with practiced speed. A few minutes later, paper in hand, she had summarized every story from the area in the 1700s she could find. Every time she found a date from the 1750s, she stopped scanning and read carefully. The silver coin Jeth had found the day before shone in her mind's eye, it's date of seventeen-something vividly standing out from the layered mud that had protected it for centuries. America hadn't even been a country then, and as she was learning, the area of Schuylkill that her grandfather's farm stood on was barely settled. In that neck of the woods, it was still pioneer land.

It wouldn't even be labeled as Schuylkill County until decades later.

A paragraph caught her eye and she paused, then read it again. Her lips mouthed the words, her fingertip flying across the page as she read. She scanned the rest of the story, her eyes going wide. She stuck her notes on the page, and snapped the book shut. She put the other two books back on their places on the shelf, then beelined it back to the main desk. The librarian looked up over her bifocals in surprise.

"I'd like to photocopy this after all." Evie handed her the book. "Just these pages." She motioned. The librarian took the book without another word, and Evie shifted from one foot to another as the copies were made.

"That will be twenty-five cents a page." The librarian gave her a curious look as she handed over the copies.

Evie handed over a dollar. "Thanks." Notes and photocopies in hand, she pushed through the doors and back into the humid summer day. She pulled her phone from her pocket and dialed Jeth.

He didn't answer.

Evie hung up and called again. This time he picked up on the second ring.

"You know, some of us work real jobs," Jeth grumbled.

"So do I! And I still manage to answer my phone."

"Look, Evie, I have a meeting in exactly thirty seconds. What do you want?"

"I found something. The name Alspach was in Pap's Bible—"

"And you couldn't leave me a message?"

"I want to go back up the mountain."

"So go."

"I want you to go along." She swallowed, refusing to mention that her fears from her youth still lingered.

He sighed. "Okay, look, we can go this weekend. Okay? Ten seconds till meeting here."

"What if Pap makes a decision before then? Don't you want to know if that old ruin was a distant ancestor's home? Don't you want to know just how much family history is on that land?"

The other end of the line went silent.

Evie stopped walking, glaring at her phone. Had he hung up on her? The screen still said they were connected. "Jeth?"

"I get off at five. Bring me dinner, and I'll meet you at Pap's at six. Better bring Pap dinner, too."

"Will do!"

"Okay, bye Evie." He hung up before she could respond.

With new energy in her step, Evie made it to her car and unlocked it, throwing her research onto the passenger seat. She was, due to her own forgetfulness of library policy, now late for work. Which wasn't the end of the world, considering she was her own boss. It just meant some serious double-timing to get things done, particularly if she wanted to get out of the shop in time to pick up dinner and make it to the farm before six.

She slapped her forehead as she settled into the driver's seat. She'd forgotten to look into Pennsylvania's geology. That section of the library was well stocked, as the librarian had even pointed out. The county did, after all, sit on one of the largest coal deposits in the world, the Mammoth Vein. Coal aside, surely there were some generalized books on the rest of the county's rock formations, not just those of the coalmine tier. Maybe she could find something that explained the gold stories.

She'd have to come back. No time now.

As Evie drove to work, her mind circled over the words she had read. Pioneer. Alspach. Son kidnapped by Indians. The part she needed Jeth and Pap's help with was pinpointing the location of the tale. To her, right now it sounded like the story could have taken place in Pap's backyard. But there were a lot of farms in the area. So much had been developed, plowed, timbered, and bulldozed for highways, would it even be possible to find such a macabre location? Why did she even want to? She reached into her pocket and turned the arrowhead over in her hand. It wasn't only the settlers' history that was at risk of being bulldozed, but the Native Americans' as well. There were always two sides to every historical tale. The urge to know the story infected her mind like an itch.

Maybe it was because, after all these years of it being just stories in books, the history was reaching out to her. That imaginary clock on discovering her family's history was ticking, just like the days the real estate company had given Pap to make a decision. An arrowhead and coin had snapped her to the realization that land doesn't realize the here and now. It absorbs all time, the footsteps of a hundred generations or more. It shifts with the movement of human growth and development, being plowed, trodden down into paths, shoveled, and piled. Land absorbs the blood of all, man and beast alike, absorbing life back into itself, that new life may come forth. Something in that little hole of dirt had made Evie realize just how much history was there on that farm, that dirt is not just dirt.

She just didn't know what to do with that knowledge yet.

Chapter 6

Jeth sipped his coffee as he stepped into his sleek Allentown office. Like a fish bowl, the entire front half was a large window overlooking his co-workers. He mused that the money would have been better spent putting in a few exterior windows, so that perhaps the group could work in sunlight instead of the sterile, eye-straining florescence that hung above their heads. The original architects likely thought natural light unnecessary. After all, the engineering crew was going to spend eight to ten hours of their day staring at the glow of computer screens anyway.

Putting his musing of natural versus artificial light aside, Jeth nudged the door to his office shut behind him. At least he had an office instead of a cubicle in the big room beyond his fish-tank wall. Instantly, his environment seemed to quiet. He needed quiet to think. He was far more productive that way.

The phone rang.

Jeth gritted his teeth and set the coffee down, flipped open his laptop and powered on the larger display monitors next to it. He reached for his company cell, holding it to his ear after swiping onto the call.

"Good Morning. Jethro Fogel-Mueller."

"Morning, Jeth. Did you get that project done yesterday?" his boss Clint asked.

"One or two things to touch up yet. I was waiting on the land survey. As of yesterday it hadn't come through."

"Client called me at seven to check on it already. How soon can you have a schematic to me?"

"As soon as the surveyors get me their report."

"Get on them. This needs to move forward."

"Let me check my emails." He waved the mouse of the desktop. For all of modern technology's progress, these monitors still never seemed to load fast enough. "I'll call you when I have it."

His boss sighed. "Alright, thanks."

They hung up without saying goodbye.

The computer loaded, and Jeth clicked into email. Eighty-two new messages. *Hello Monday.* He gritted his teeth and started clicking through, downing his cooling coffee as he went. It was amazing what petty things his co-workers needed to ask him about. Half of the emails were unnecessary handholding. He typed short, un-flourished replies when he had to, deleting the rest, which mostly consisted of company-wide memos that no one actually read. He scanned the latest one that was a reminder to review their annual benefits package. As if the three letters in the mail weren't enough of a reminder.

He starred that email. Maybe he did need the reminder since he still hadn't updated and reviewed his account. Surely health insurance had gone up again. Better find out just how much. Hopefully his salary would see a proportionate increase.

Finally, Jeth found the email from the surveyor, sent at 11:48 the night before. With relief he skimmed the attachments. All the information he needed to finish his part of the project was there. Without so much as a thank you reply to the poor soul who'd pulled a late night to get it done, he settled into his own CAD programs, entering in this vital information. By eleven-thirty-eight, he was calling his boss as he clicked send.

"About time," Clint grumbled. Jeth heard the rustle of papers in the background. "Look, we have a situation with that new account, RimForce. I need you to prioritize that."

"You told me yesterday to prioritize the Magen project. Who's going to take over that?"

"Just put it on hold. I'll figure it out." He hung up.

Jeth stretched and looked at the clock. Almost time to take a lunch break. He scanned through two dozen new emails, then clicked open the RimForce, Inc. project file. It was a doozy. It was going to take months to work through the data they needed to start building the mega-warehouse they were planning. He logged out and clicked on his screen savers just as a pretty blonde popped her head into his office.

She glanced over her shoulder and slipped in, shutting the door behind her with a sneakiness that he couldn't help but find amusing, considering the glass window revealed all. There were no curtains or blinds. Her stilettos clacked on the linoleum as she stepped over to him. His eyes caught on her blouse, that maybe should have been buttoned just one hole higher.

He leaned back and swiveled in his chair to face her, a smile creeping onto his lips.

She leaned on his desk and bit her lip. Her eyes sparkled with a hidden laugh.

"Lunch?" she asked.

"Absolutely." He pushed to his feet and pocketed his two cell phones, one work and one personal. He glanced at his watch. "I've got to be back in thirty though."

She rolled her eyes. "Me, too. Clint call you about the RimForce project?" She followed him out of the office.

"Yup. Going to have some long hours on that one." He held the door for her as they exited the building.

They walked a short way down the street where finally Jeth felt they were far enough away from the office that he could pull her into his arms and kiss her. She tasted of mint and smelled of expensive perfume.

Hand in hand, they continued down to their favorite sandwich shop. "Missed you last night. How did your family thing go?" she asked.

He smiled. "My cousin is crazy. But we had a good time."

"That's all you're going to tell me? Did you find anything?"

"Sienna, I invited you to come along. You could have witnessed it firsthand." He held open the door to the sandwich shop.

"I know. Hiking isn't really my thing."

He stifled a laugh, running his eyes from her manicured nails to her four-inch stilettos. He got stuck at her legs. She had awesome legs.

"You do realize that hiking is the same as what you do on your treadmill—you just actually go places when you do it—right?"

She stuck her nose in the air with a mock sniff of disdain. "There are no bugs near my treadmill."

They both laughed. But Jeth felt that little wall of hesitation towards his girlfriend pull at the pit of his stomach. At times he felt like he lived a dual life. When he did city-stuff with Sienna, they were a dream couple and had a blast. But his life back at Pap's farm was as foreign to her as a new country. Hanging with his cousin had brought all that to light, and his reservations had returned regarding the direction of his relationship with Sienna.

He had yet to admit to her that he hunted.

"I'll have the avocado grapefruit salad on the vegan wrap," she told the server.

His smile faded slightly. They had been dating discreetly for almost a year. They knew each other's favorite colors, movies, and ice cream flavors. It wasn't like he was keeping his hobbies a secret. He just hadn't told her yet. "Philly cheese steak. Extra sauce, please," he ordered. He gave her a sideways glance. She didn't even bat an eye at his order. Never did. Maybe it was time he showed her his other life and found out if she liked that Jeth, too.

A buzz vibrated from Jeth's pocket, and he sighed. Personal phone. Evie.

He shot Sienna an apologetic glance and answered the call. "What's up now, cuz?"

"Jeth, I can't meet you tonight. Can we reschedule for tomorrow? I messed up an order and now the only copies of this book in stock are in New Jersey. And the customer needs them tomorrow for her book club. But I'm bursting with what I found. Can I tell you now?" Evie gushed.

"How about tomorrow?"

"So in the 1700s there was this family—"

"Evie, I'm on my lunch break right now. With Sienna. Can I call you later?"

There was a pause on the other end of the line. "I seriously need to tell you what I found. Can I call you tonight on my way home?"

"Sienna and I have plans."

"You said this morning you could meet me tonight. You couldn't possibly have made plans yet!"

"Just made the new plan." Jeth winked at Sienna. She smiled back, a twinkle in her eye.

"Then I'll call you after you're done work. Before your dinner plans. What time you done?"

Jeth calculated the length of time he'd need for the RimForce project. He'd probably have to do overtime to even get it started. "I dunno, Evie. Probably late." He could feel her disappointment through the phone. "I'll call when I'm done, okay?" He swiped his credit card at the cash register and paid for his and Sienna's lunch. Sienna took the bags.

"Fine. Don't forget." Evie seemed placated.

"Bye, Evie."

"Bye, Jeth."

He hung up.

Sienna's eyes twinkled with laughter. "Your cousin seems a bit...enthusiastic."

"She's doing some research on what we found yesterday. Going a little overboard with it, as usual. Apparently she found something *super* exciting." He rolled his eyes with sarcasm.

"What did you find? You never answered me."

"An arrowhead and a very old coin."

"That *is* interesting." She said it sincerely enough, but her gaze was locked out the window as raindrops spattered the panes of the sandwich shop. She looked longingly at an empty table, then outside, then at the clock on her phone. "I hate to be like this, but—"

Jeth sighed. "I'm with you." It was going to be lunch at his desk today. They both had too much work to do. He gave her a quick peck on the lips before they ducked through the door. "I'll see you tonight."

They ran through the rain back to the office, Sienna's heels clicking on the pavement as Jeth's collar got a little tight around his neck.

Chapter 7

T he storm seemed to have risen from nowhere. Ten minutes earlier, before Evie had pulled onto the interstate, the sun had been shining from a blue sky. She hadn't noticed if there were clouds. There was too much on her mind. There sure were clouds now, and they were deluging rain in sheets that her autobahn-approved windshield wipers were supposed to be able to handle. They weren't. She had slowed to forty-five and slammed her thumb into her four-way flashers. She hated this highway, but it was the fastest, most direct way to get into mid-New Jersey. With the time she'd wasted in the library this morning, she needed fast and direct to stay on schedule and pick up the inventory she needed.

She was still annoyed with her ordering mistake. Where was her head lately? Was it the stress of the pending sale of Pap's farm or something else entirely? Like Jeth's casual mention of her ex-boyfriend?

An SUV zoomed past her in the left lane, throwing a sheet of water onto the already overwhelmed wiper blades. Evie was momentarily blinded. She squinted and slowed down a little more, leaning forward over the steering wheel, as if pressing her nose to the glass would somehow improve visibility.

"If I can't see, how the heck can you?" she mumbled at the passing driver whose tail-lights faded around the next bend.

She leaned back again, knuckles white on the wheel, and debated pulling over. As if in answer, the rain came harder, the shoulder of the road no longer visible. She stiffened a little as she noted a car that had pulled over, wishing they had pulled over further. She glanced in her rearview mirror to check if the road was clear, then merged to her left, leaving the little car more room. It was bad when it was raining too hard to stop. She didn't want to be rear-ended.

The view from the left lane seemed to give her a clearer view around the next turn, and she leaned towards her dash again, squinting at the hint of brake lights ahead. She merged right. A few seconds later she realized the lights ahead had not moved, and now were significantly closer. Even at forty miles an hour, she was closing on them fast, too fast. She slowed significantly, glancing in the rearview mirror. The bright headlights of a tractor trailer were in the distance behind her, plenty far away. She wished she had a second set of flashers. Then the scene before her appeared through the rain. Cars were askew all over the place, and at least two tractor trailers seemed to have slid off the highway. She pulled to a stop on the right side of the road, heart pounding. As she tried to process what she was looking at, someone started pounding on her passenger side window. She looked at the person in panic, her first instinct suspecting it was a thief. Odd place for a thief.

"Get Out! Get out!" the person's screams finally reached her ears. The man was looking behind her, and Evie saw the tractor trailer she'd seen in the distance was closing in, and did not seem to be slowing. She unbuckled her seatbelt, grabbed her phone and keys, and threw open the door. The man who'd banged on her window met her at the back of the car and grabbed her arm. She didn't protest as he propelled them both up the hill with superhuman strength. Her legs pumped as fast as she could, and still the screech of metal tearing seemed far too close.

She didn't look back. She couldn't, not with the man's iron grip on her arm. At the crest of the embankment, he released her. She barely got a look at his grey-bearded face, reminiscent of a biker Santa Claus, before he was gone again. Evie gasped for air as the hands of strangers touched her shoulders.

"Are you all right?"

"That was close."

A half dozen men and women and a girl who looked to be about ten stood around her. Only then did Evie process what had happened, and she turned back to face the highway.

Her car was gone. In its place was a huge semi that was embedded in the back of another semi. Another SUV came sliding into a brake-squealing stop, just not soon enough to prevent slamming sideways into the back of the semi. A little car came onto the scene slowly, like Evie had, then gunned the engine and pulled far up onto the bank of the road, off-roading past the two impacted semis before stopping. Evie looked beyond where her car had been. The highway was a jumble of cars and trucks stretching into the distance, the steady rain obscuring the end of the pile-up.

She started shaking.

Another semi came barreling up, this one mercifully slower than the last. He came to a stop mere feet from the sideways car. Seconds later, another truck slammed into his backside. It was like a bad car crash movie that wouldn't end. The little girl with the group started to cry, and her mother pulled her tight into her arms, shielding her eyes.

As another vehicle could be seen in the distance, Evie registered that the car smashed sideways into the tractor trailer still had people inside.

"Has anyone called 911?" she asked in a shaky voice.

"Yes. We all did," someone replied.

Evie took a step forward towards the wreck. Someone caught the back of her shirt.

"Don't. It's too dangerous."

"But if they need help to get out..."

"You'll only get in the way."

As if to accentuate his point, another car screeched to a stop. That car's occupant immediately exited and barreled up the embankment towards them. He hadn't quite reached the group when another tractor trailer smashed into it, pushing it spinning off the highway into a ditch before continuing forward to obliterate half of the trailer in front of it.

The man that had run turned and watched, issuing a long line of expletives. The mother covered her daughter's ears.

"Where is the man that saved *me*?" Evie looked around for Biker Santa. He wasn't in the group.

The downpour eased to a drizzle, and the pile-up seemed to stop piling as vehicles managed to stop with plenty of space between each other. Evie shivered as the adrenaline started to wear off, realizing she was soaked through to her underwear. She checked her phone. Only four minutes had elapsed since she'd last looked at the clock in her car. That quickly, everything had changed.

The acrid smell of smoke reached her nose, and someone pointed to the billowing cloud a few hundred yards down the highway, the tangled mess of vehicles stretching for over half a mile. The black mass of that smoke column contrasted to the sun that suddenly burst out from behind the clouds. Evie didn't think about that irony. Her feet carried her down the hill on their own volition. Where was her rescuer? Where was her car?

The other survivors, except for the mom and daughter and an old man with a cane, came down with her. Some ran to vehicles, peering into windows. Others took the leisurely pace of the shocked. The tractor-trailer driver emerged from the truck that had stopped safely. He surveyed the back of his rig with his hands pulling his hair in dismay. Evie approached the car that had skidded sideways. The male driver was slumped against the side of the door. Her blood ran cold. Was he dead? She stared at the rise and fall of his chest through the passenger window. No, not dead. A sound issued from the back seat. She pressed her hands to the glass. Two children sat safely strapped into their car seats. They could not be more than five years old. A little boy was screaming at the top of his lungs. She tried the door. Locked.

The truck driver appeared at her side. "Driver okay?"

"I don't think so. But there's two kids in the back. I can't open the door."

He uttered a list of expletives under his breath, looking at the distance between his bumper and the car. His eyes went wider, and he was visibly shaken. "I almost killed them."

Evie grabbed his elbow. "Hey, they're going to be okay. You have anything—crowbar maybe—we can get them out with?"

He seemed to rally with that. "Yeah, sure." He disappeared into his truck, returning with a massive wrench. Eyeing the vehicle, he said, "I'm going to break the front passenger window. I don't want to shower the kids with glass. Stand back."

Evie stepped away, listening to the sound of breaking glass as her eyes flicked across the carnage in front of her. An ominous smell came from the back of the parked rig. She squatted and looked under the tires. Flames dropped out of the truck that was embedded in the trailer, hissing as they hit the wet pavement and sputtered out.

"We've got to hurry." Evie stepped back to the trucker as he smashed in the last of the window with his elbow.

He reached inside the door and pressed the lock. The panic alarm came to life, but the locks didn't budge. "That's technology for you," he grumbled. He eyed Evie. "You think you can crawl in there and get those kids out? I certainly can't."

She didn't hesitate, swinging her leg over the broken window, wary of the glass. For once it was good to be petite. The kids looked at her with wide, wet eyes, their noses dripping mucus. "Hey, it's okay. We're going to get you out of here." She unbuckled the smaller child, and pulled her from her car seat. The little girl miraculously didn't protest, and wrapped her arms around Evie's neck. Evie passed her out the window to the trucker.

"What kind of cargo are you hauling, anyway?" Evie asked as casually as she could, reaching for the little boy. He seemed eager to help her, shrugging off his harness once she unsnapped it and reaching for her with fat little arms. Her back protested at the odd angle as she passed him to the trucker, who then had a child on either hip.

"Potatoes. Why?"

"Just so it's not anything explosive." Evie turned her attention to the driver. He still breathed. She undid his seatbelt, hands and eyes inspecting him for injury. When she touched his head, she could feel the bump on the side of it. He blinked, his eyes unfocused. "Thank God," she whispered.

She glanced back to the trucker. The mom that had stayed up on the hill appeared at his side, taking the little girl into her arms. The little boy was placed on his feet, his hand firmly in the mother's. She guided them up the hill, and the trucker turned his attention back to the car.

"I can't move him," Evie groaned. Getting him out via his door was impossible. The entire side of the car was pressed against the back of the trailer he'd hit.

The trucker glanced behind them. On impulse, he took off his belt. "Here. Can you get this around his chest? Or shoulder and neck?" The trucker had a substantial girth, and the belt fit the driver with room to spare.

"Now what?"

"Support his head, and I'll pull." The trucker reached over Evie, grasped the belt, and the man started to slide across Evie's lap towards the broken window. She carefully supported his neck, praying he had no spinal injuries.

"Watch the glass!"

"He's going to get a few cuts. Better that than burn alive," the trucker grumbled. With a heave, the man's head and torso passed Evie. He seemed to be regaining consciousness, helping to steady himself on the window frame as he passed. By the time the trucker got him through window, he was able to sit up on his own.

Evie crawled out the window behind him.

The trucker looped the man's arm over his neck. Evie supported his waist. Together, they guided the driver on his wobbly legs away from the vehicle. Evie stopped at the bottom of the hill.

"We aren't far enough away yet." The trucker tipped his head, now beaded with sweat, towards a propane truck only a few yards away from the fire. That driver was trying to maneuver his truck further from the flames with only marginal success. He got stuck in the grassy median, then abandoned the truck to its fate.

Evie swallowed and took more of the man's weight as they worked their way up the embankment. She motioned to the mom, who had now three kids in hand, and the group of them coalesced with another group of survivors further from the fires. The group was steadily growing. Some people sat in the wet grass, stunned. Others nursed minor wounds. Still others were frantically speaking into their cell phones.

Evie scanned the crowd for her rescuer. He was nowhere to be seen.

The trucker set the father down in the grass, and he sat there, head in his hands. His two kids rushed into his arms, and after a moment of confusion, the man burst into tears.

"Oh, you're okay, you're okay. Thank God you're okay." The three of them cried in relief.

Someone appeared with an ice pack and placed it on the father's head. Evie and the trucker stepped away, surveying the scene around them.

A woman in a business suit stepped up to the trucker. "I've called 911 and no one seems to know when authorities are going to arrive or how we're going to get out of this mess. Did you hear anything over the CB?"

The trucker nodded. "They know. They're on their way. But that's all I heard—before—" He shook his head, emotion welling up.

The woman laid a hand on his arm, giving him a reassuring smile.

On cue, sirens could be heard in the distance.

Evie again looked for her car. Now that they had moved further down the highway, it should be visible. Two men approached, one supporting the other, who was bawling his eyes out. Evie's trucker approached them, offering support, which was pushed away as the man sat down and put his face to his knees, his shoulders shaking.

The trucker returned to Evie as if they were old friends. He had tears in his eyes.

Evie looked to him in confusion. The man didn't appear to be hurt. "Did he lose someone?"

The trucker sadly shook his head. He pointed to the truck that the car they'd rescued the family from had been pinned against. The front end was partially embedded in the truck in front of it, the mangled remains of a little blue car squashed between the two. There was no way anyone could have survived that crushing blow. "He killed someone."

Evie's vision sparkled with stars and she felt her legs give way until she sat on the ground. Her pants instantly soaked through with mud, but she hardly registered it. The trucker knelt beside her, a hand on her shoulder in concern. She gasped for a breath as emotion overwhelmed her. Her eyes stayed locked on the mangled blue car.

"Tell him—" she gasped. "Tell him—he didn't kill anyone. I—I'm alive." Her shoulders shook, and then the wave of tension broke in a flooding wave of sobs that left her throat raw and sides heaving.

Chapter 8

Jeth glanced at the caller ID and then the clock. It was only 4:53 p.m. Evie had no patience. He silenced the call and went back to typing. The phone rang again. He slammed the enter key on the keyboard and picked up the phone, swiping to answer.

"Evie, I'm not quite done yet," he grumbled, taking no effort to disguise the annoyance in his voice.

There was a sniff at the end of the line. Then a moment of silence. "Jeth, I was in an accident."

He straightened. "You okay?"

"I'm fine. My car isn't. Look, I'm only about ten minutes from your office. They bussed us to the firehouse. Can you pick me up? My car is totaled."

Jeth's brow furrowed in confusion. "Bussed you? Who is us?"

Evie sighed. "Look, turn on the radio on your way here. I called my mom, but she's still two hours away at that conference. I've been in soaking wet clothes for half the day. It's ten minutes away, Jeth. Please?"

His confusion only worsened, but Jeth was already pulling on his coat and turning off monitors. He pocketed his work phone and slipped his laptop into his briefcase. "Where are you?"

"County Firehouse off exit 49. Thanks, Jeth."

"I'll be right there, Evie."

He popped into Sienna's office long enough to utter "family emergency" and promise to call her as soon as he could assess the situation. She waved him out with genuine concern. When he was in his car, he tuned the radio to the local channel. Nothing but music. So it shouldn't be that big of a deal. The song ended as he pulled onto the highway, two exits away from Evie, and noted the flashing highway advisory sign.

Interstate Closed Ahead. Tune to 1530AM.

Jeth obeyed, instantly tightening his grip on the wheel. The AM station's tinny static filled the car, then the message played.

"Highway 78 closed between mile marker 40 and 29. Avoid the area if possible."

What could have caused an eleven-mile stretch of highway to be closed? An ambulance blasted past him going eastbound in the opposite lane. Eerily, there was not much traffic on that side of the highway at all. What had Evie been involved in? He accelerated, only to be forced to slow as the traffic became congested. The closure ahead was backing up the highway ten miles away.

It took far longer than ten minutes, but Jeth made it to the fire station. Firefighters were out directing traffic off the exit ramp. He rolled down his window as one tried to flag him past the station entirely.

"I'm picking up my cousin. This the place?"

The fireman nodded and waved him through. Another directed him to a parking spot. The place was swarming with cars and a steady stream of buses. Buses from where? Some people were running from their cars as soon as they parked, their eyes wide with fear. Jeth saw more than a few tearful reunions taking place at the firehouse doorway.

He closed his car door with a thud, then stood for a moment, taking it all in. He called Evie as he approached the building.

"Where are you?" Jeth asked.

"I'll meet you outside," she said. "It's chaos in here. They just brought another bus. I can't believe they're still triaging people." She hung up, and Jeth found a place to wait underneath the flag pole.

When she appeared, Evie looked like a drowned rat, her hair a tangled mess, her makeup smeared, and her clothes still damp. Jeth considered making some kind of joke, but thought better of it when he saw her face. It held none of the excitement he'd heard emanating from her earlier that afternoon. She scanned the crowd, looking for him, then beelined it toward him.

He gave her a light hug when she got there. "What on earth is going on?"

"Didn't you see the news?"

"It wasn't on the radio. Just the highway closure."

"Thanks for coming." Evie glanced around for his car, and once she spotted it, she led the way there.

Jeth followed wordlessly, holding the door for her to slide into the passenger seat. When he had secured himself into his own seat, he turned to her. "You sure you're okay? Want to get checked at the hospital?"

Evie waved him off. "I'm totally fine. I wasn't even in the wreck. And they just checked me out here anyway."

"But you said your car was totaled."

"It was." She took a deep breath. "It was a pile-up, Jeth. It was horrible. They haven't said how many vehicles or how many are injured—or dead—but it was horrible. A solid half-mile of metal, if not more. All of our cars are still in the wreckage. It's going to take days to clear. And there were fires…" Her voice trailed away. "Can you just take me home? I'm exhausted. Thanks for coming."

"Sure, cuz." He started the engine. Normally he would take the highway, but since it was closed, they had a long drive the back way home, up and over the Blue Mountain. It added at least twenty minutes to the hour-long trip. He had to call Sienna and tell her he'd have to reschedule. He handed his phone to Evie.

"Can you call Sienna for me?"

"Sure—oh, Jeth. I'm sorry. You have a date! I forgot." Evie punched the speed dial for Sienna with a little more force than necessary. "Should I just call an Uber?"

"It's fine," Jeth mumbled. He turned onto a road that led them through the backside of the warehouse district. Each one seemed larger than the last, their parking lots full of tractor trailers, acres and acres of asphalt surrounding them, and manicured water retention ponds with tiny, newly planted trees breaking up the monotonous concrete and steel.

Evie spoke with Sienna, apologizing profusely. Jeth could hear Sienna's voice. "I just saw on the news. Thank heaven you're all right..."

The two women wrapped up the conversation with Evie promising to give Jeth some great date ideas, and she hung up.

Evie slumped back in her seat, her head against the car window. Suddenly she cursed.

Jeth's eyes went wide.

"I know better than to drive that section of road! I know it! It's a freakin' death corridor. Too many trucks. And then we wonder why." She gestured angrily at the warehouses surrounding them. "They just keep building. Look at them. All of them. Warehouse. Warehouse. Warehouse. Brand new. And thousands of trucks. And the trucks use our highway." She slumped back again, biting her lip, fighting back tears.

Jeth was quiet a moment. "They're working on expanding the highway. It will help."

"Expand it where?" Evie's voice took on a dejected tone. Her question was the very one Jeth's company was being paid to solve. Where? How? How far? When? But he caught her greater point. It was farms like Pap's that were going to lose acreage. Sure, the highway may only need a few hundred feet to be safer, but the cost of farmland led to a whole set of problems he didn't feel like thinking about when he wasn't in the office. It meant farms next to highways became far more valuable as warehouse real estate than planting real estate.

The thought settled deep in the pit of his stomach.

He tried to shift the conversation with Evie. "What did you want to tell me this afternoon?"

"Oh, before my car was eaten by a tractor trailer?" She huffed. "All my research—and my purse—are still in there. I'm going to have to cancel all my credit cards tonight." She sighed and rubbed her forehead. "In the library there was a story about the Alspach family. That name was in Pap's bible."

"Okay."

"They were among the first white settlers in our area. 1750s. Then they were killed by Indians. The father, Peter, came home to his wife and son murdered. His youngest son, who was only a baby, was taken captive."

"That's wild."

Evie nodded, some of her old enthusiasm returning. "The youngest son was eventually traded back to the family by an Indian chief, but so many years had passed that he no longer spoke his native language."

There was a long pause.

"And then what?" Jeth prompted.

Evie shrugged. "That's all it said. Peter later remarried. He had more children, who might be our ancestors. He rebuilt, and he stayed in the area. The Indians lost their land when the French and Indian War came to a close and the treaty lines were drawn. The fighting ceased, at least around here. And the rest is history."

"And how is this important?"

"What if the site of the massacre is the old cabin on Pap's farm?"

Jeth felt his eyebrows shoot up. "Now that's an interesting philosophy. How on earth would we prove it?"

"We found the coin from that era. And an arrowhead."

Jeth drummed his fingers on the steering wheel. "It does sound likely. Wow, so that would mean the farm has actually been in our family..." He counted in his head.

"For two hundred seventy years." Evie watched him, her expression serious. "You see now why I needed to talk to you? How can we possibly let this end with us?"

"Evie..."

She waved him off, turning her attention back out the window. "I know. Money talks. I'm just saying, I think we should have a say."

Jeth thought hard as he drove, now winding via Route 309 over the Blue Mountain. The leaves were just starting to change, a splash of yellow here and there. "I wonder if my mom told my sister the farm is for sale."

"I'm sure she did. But Marissa's out of state, finishing her doctorate. It's not like she's going to take up farming."

Jeth glanced at the clock. "Well, since my date with Sienna is canceled, want to go back up the mountain and look for more clues?"

Evie shook her head. "Sorry, Jeth. I really just want to get home."

There was a long moment of silence, then Jeth voiced, "Just imagine what kind of native history is in that soil, too. If you found an arrowhead, it means they passed through there. What if there are more?"

"I suspect there are."

Jeth nudged her. "You're going to keep me in suspense? Leave me alone with all these treasure-hunting questions? First gold, now historical artifacts."

"Now you know how it feels." She smiled, and he was relieved to see it.

A little over an hour later he pulled into Evie's driveway.

"I'll give you a call on the weekend," Evie mumbled. "I'm going to be playing catch-up for days after this." She slapped a hand to her forehead. "Oh, I didn't get the books! Now I have to call the customer and explain. Ugh...."

"Hey, at least you're okay. The rest is a small matter."

"Yeah. I know." Evie met his eye, one hand still on the door. "Hey, seriously, Jeth. Thanks for getting me home. Sorry I messed up your night."

"That's what family's for, right?"

"Well, I appreciate it. I'll make it up to Sienna."

"Goodnight, Evie. Go take care of yourself."

She closed the door with a last weak smile. He waited until she had unlocked her front door and gone inside before he pulled out of the drive. His mind raced as he drove to his own house on the opposite end of the little town.

When he flicked on the evening news that night, his mouth dropped open. The carnage of the highway was shown by drone footage. Semis and cars and SUVs were indiscernible from each other in a long train of metal. Someone had posted a phone video of the scene, cars and trucks smashing into each other with the screech of metal. It had already gone viral. Jeth wondered how Evie had survived without so much as a scratch. The TV said several were dead, and crews planned on working through the night to clear the wreckage.

As the newscaster changed subjects, he turned to YouTube, watching the viral clip in its entirety. The poor visibility from the freak storm was at fault, they said. But as he counted the trucks, calculating in his head what volume of traffic the highway was engineered to carry, the answer became more ominous.

RimForce was paying his engineering firm millions for a reason. The numbers didn't add up. The volume of traffic was expanding at a rate faster than the highway expansion project could handle. The problem at hand was going to require the attention of more than just his little firm to address. It was going to involve government on every level. It was going to involve a whole new infrastructure.

Chapter 9

E vie dropped her keys on the side table next to yesterday's unopened packages. Pity she hadn't just ordered the customer's books correctly the first time. Then she wouldn't have been on the highway. She would not have wasted a day. And her car. Oh, her poor car! She couldn't think about the consequences if Biker Santa had not told her to get out. She had not seen him, even at the firehouse. She didn't even know his name.

The packages would wait. She needed a shower. But the businesswoman in her forced her to sit down to make the necessary calls to cancel her credit cards and apologize to her client, who thankfully was more than understanding. A quick text to her best friends Jayelle and Beth about the accident solicited the appropriate emojis and a promise to get together soon.

Exhausted, she set the phone down and stripped out of her clothes as she walked, dropping them unceremoniously in a pile as she set the shower temp on high, letting the cloud of steam fill the bathroom before stepping into the heat. She no longer felt like crying. Her tears were dry. She'd spent them all sitting in the mud next to a trucker.

In the chaos after emergency crews had arrived, she hadn't asked his name. Biker Santa, the trucker—she hadn't gotten anyone's names. They were just a string of lives spun together at one solitary point before being pulled back apart, like a blanket on a loom with only one or two rows pressed together, the loose ends held apart by different points of destination. Her father, God rest his soul, would have scolded her for not being more forthcoming.

Evie let the water run over her for a long time, then finally added soap and rinsed clean. She wrapped a towel around her head, pulled on her comfiest clothes, and threw a TV dinner in the microwave. As the microwave whirred into action, she went back to her packages, slitting the tape on each with a key.

The contents gave her at least *some* pleasure. Two new books, historical romances with very handsome men on the covers, would hopefully distract her for the night. The other box contained a few months' worth of cleaning supplies, compliments of a particularly addicting company. How good of a marketing campaign did they have if you got excited about cleaning supplies? She held the paper packaging material in her hands, wondering just how green the company really was being if they needed to use such big boxes with so much packing material for a little order. She made a mental note to try to group her orders together more. Then she changed her mind on that as she piled up the freebies that were included with her dish soap and bamboo toilet paper. It was like one of those mail-order subscription clubs for grownups.

The microwave beeped, and she fetched her dinner, pulled her feet under a blanket on the couch, and thumbed open one of the new books. A few hours later, sufficiently engrossed in the book and having almost forgotten the day, a knock sounded at the door. Evie sighed and went to check who it was. The clock said it was only eight. It felt like midnight.

Her mom didn't wait for Evie to answer, but came right in. "Knock, knock!" she called.

"Mom. What are you doing here?"

Wordlessly her mother pulled her into a bone-crushing hug. "I just needed to see you." Her voice shook with emotion. Obviously, she'd watched the news. Evie wrapped her arms around her mother's waist and held tight, breathing in the familiar scent of Dove soap. After a long moment, they pulled apart.

"You okay, sweetheart?" her mom asked.

"Just tired."

"I bet. Did you eat?"

Evie gestured towards the remnants of her TV dinner.

"Good." Mom pulled her back into a hug. "I saw it on the TV. I'm so glad you are okay." When she released Evie, she wiped her eyes. "Here. I brought you cookies." She reached to the side table by the door and produced a white box.

Evie laughed. Cookies had always been her mother's cure for emotional turmoil. She led the way into the kitchen, set out plates for both of them, and they settled on the bar stools at the island. The frosted sugar cookies were amazing, and they both reached for seconds, licking icing from their fingers.

"Where'd you get these from?"

"Little bakery in Hershey. I wish they were closer."

"Uhm-hmm." Evie agreed. As she bit into cookie number three—hey, it had been a long day—she thought back to the morning. "Hey, Mom, did you know there's an old cabin—well, the ruins of an old cabin—at Pap's?"

Mom laughed. "Of course. Used to camp up there all the time when we were kids. Is it still there?"

"Not much of it. So was it you and Aunt Maggie that used it as a party spot?"

"I have no idea what you're talking about." She smiled, a little glint of mirth in her eyes.

"Uh-huh." Evie rolled her eyes. She got up and went into the bathroom, rooting through her wet pile of clothes to locate the arrowhead. Thankfully, despite the day's events, it was still there in her pocket. "Check this out." She put it on the table in front of her mother.

"You found this up there?"

"Yup. With a coin Jeth says is from the 1750s. Pulled them out of the same hole."

"Really?"

"Isn't it neat?"

"You betcha." Her mother spun the arrowhead in her hand. "We never found anything that interesting up there. Just a lot of nails. Always nails. Usually under my sleeping bag."

"We found those, too." Evie hesitated. "Mom, do you think there are more artifacts up there?"

Mom was silent for a moment, then firmly shook her head. "Evie, you shouldn't go poking around up there. That kind of thing gets the authorities involved, and with the sale of the farm pending, you don't want anything to slow up that process."

It wasn't the response Evie was hoping for, and her excitement faltered. "Don't you want to know if there was an Indian settlement there? I think it's interesting."

"So might people that don't want Pap to sell, sweetheart. That is why we are trying to keep everything quiet."

"How is it going to be a secret if they build a giant warehouse in the middle of his farm?"

"By then it won't matter to Pap. He'll have his retirement. If the community finds out before the sale goes through, they could protest at the township meetings, harass Pap—at the very least, they are going to try to sway his decision."

Evie frowned. She could picture the uproar among the neighbors if Pap did sell, and they found out a warehouse was going in next door. "Don't you think they all should have a say in the process as well?"

"They do, and they will. After the developer hands over ten million dollars. Then the legalities are their problem. We don't want them to be ours."

Evie reached for another cookie. "You think he'll actually sell?"

"Well, I hope so. He deserves his retirement after all those years. I told him to buy a nice place on a lake. He could fish all day. The homeowner's association would help him keep up with the maintenance. And lord knows, at that price, it will leave you and your cousins sitting just fine."

"You aren't sad about the farm...not being a farm?"

Mom sighed. "I have mixed feelings about it. But the family hasn't seen an offer like this in two hundred years. And it's not like anyone in the family is going to take up farming anytime soon." She laughed.

Evie didn't laugh with her.

"You okay, honey?" Her mom rubbed her shoulder.

"Just tired. It's been a long day."

Thankfully, her mother took the hint and stood to leave. "I bet. You rest up. Take care of yourself." She gave her a firm hug. "I'm glad you are okay."

"Thanks for the cookies. And for stopping."

Her mom laid a hand on her cheek. "Any time you need me, call." She closed the box of cookies and gave Evie a warning look. "And don't eat this entire box in one night."

Evie smiled as best as she could, followed her mother to the door, and locked it behind her. She leaned back against the frame. She had needed her today, and as usual she had been occupied. Which was perfectly understandable. The cookies proved she was worried and cared, but it still had fallen on Jeth to rescue her today. She couldn't help but wonder if Pap felt the same. Two daughters, too successful and distracted by their own lives to step into the management of the family farm. Three grandchildren, one of whom lived in California, also very successful and wrapped in their own lives. Evie thought back to a few summers ago, when Pap had asked her to help unload hay. She'd pulled the "I'm a girl" card without shame. Who wanted to unload a thousand fifty-pound bales of hay in ninety-five-degree heat? That was the definition of a man's work.

She wondered if Jeth had gone to help. Had he been busy at the office?

She turned back to her couch and her book. Her tea was still slightly warm, and the chamomile was finally taking effect. She thumbed through the pages. Somewhere in there was a handsome knight that didn't abandon his lady to the dragon.

She wondered what had happened to his family's farm.

Chapter 10

By the time Jeth had finished work the next day, he had a roaring headache. Whatever nerves were still connected in his brain felt like they'd been fried like eggs on a nice hot pavement. Not on a skillet, like you would clean eggs. On a dirty, rough pavement. That was his day. And it all stemmed from the RimForce project.

He threw his briefcase on the front seat of his truck, debating whether he wanted to stop and get a beer at the local bar or just go straight home to crash on the couch.

"Hey, Jeth…" Sienna appeared from two cars away. "I told them they need to hire on a temp for our team. This is ridiculous with the timeline they're pushing."

"I agree. They won't though."

"Probably not. You want to get a drink?"

Jeth thought about that beer, but for some reason it didn't sound appetizing anymore. The disadvantage to covertly dating a co-worker was that work had a tendency to creep into conversation, particularly when the projects were as overwhelming as this one.

"I think I'm just going to head home. I want to check up on Pap. I think he's trying to get the oats harvested this week." He rubbed his temples. "Maybe a little fresh air will magically give me the solution to this project. You want to come? We could go for a walk."

He could see Sienna debating as she bit her lip. She glanced at her watch. "Uh, I think I'm going to get some stuff done at home, too. Then I'll be able to think more clearly. You have plans tomorrow?"

"I owe you a date." He pulled her hips into him, not caring if the parking lot cameras caught him.

"You do." She smiled, then kissed him lightly on the lips. "Tomorrow, then."

The whole drive back to Schuylkill County, Jeth thought about the work project and Sienna. His mind spun in circles. He took the highway, sticking to the hammer lane to get around the endless stream of tractor trailers. Thankfully the accident from the day before had been cleared. The only sign it had ever happened were the skid marks in the turf along the highway. As he turned up Route 61 through the Port Clinton pass, a cleft in the Blue Mountain, he mused at how his ancestors must have traveled the same route. It was a hassle, even now in the twenty-first century, to cross the mountain. It added time. The roads closed in the winter if the snowstorms got too bad. He could not imagine trying to cross that steep hillside with a horse and wagon, even in good weather.

This little cleft in the mountain, cut by the Schuylkill River, had been the main artery through the wilderness separating the wide, fertile valleys to the south and the wild, hilly country to the north for centuries.

He let his eyes run up the bank to his right, knowing from experience that the Appalachian Trail climbed that seventy-degree slope in a calf-burning ascent.

It had to be hard to be a settler in these hills. Once you crossed the mountain, where were you? You were lost in a wilderness that the Indians still claimed as their own. Was it theirs? He'd have to ask Evie if she'd researched that much. The whole story seemed fantastical to him. And yet he'd passed the historical markers for the forts in Auburn and on Route 309 that had been in use during those early years of settlement as protection for the settlers. There must have been a need for forts to warrant building them.

Soon Jeth pulled into Pap's driveway. It wasn't hard to find him. The big red Massey-Ferguson tractor was parked in the barnyard, and Pap was on his hands and knees under it, twisting to look up into the belly of the old machine.

"Uh, hi, Pap. Whatcha doin'?" Jeth leaned forward and peered under the tractor.

"Hey, kid." Pap back crawled out from under the tractor, pulled a rag from his pocket, and wiped off his greasy hands. He pushed off his knees very slowly, wobbling with them bent for only a moment before straightening and turning to Jeth. "Just changed the oil so she's ready to go for harvesting tomorrow." Jeth knew Miss Massey pulled the grain wagon, while the combine did the actual cutting of the oats. "What are you up to?"

"Wanted to see if you needed help. I guess not if you aren't combining yet."

"No, I'm about done for the day. Could use the help when we bale straw, though. Gotta get it done before the rain Saturday. You gonna be around?"

Jeth shrugged. He couldn't blow off Sienna yet again. The rest depended on the RimForce project. "Not tomorrow at least. Got a hot date, Pap."

"Oh my, my." Pap's face lit up. "Did I meet this one yet?"

"Not yet." Jeth smiled.

"Well, when you going to bring her by? I hope I don't have to wait till Christmas."

Jeth laughed. "Maybe Thanksgiving."

Pap gave him a light punch on the arm. The two of them picked up Pap's tools and Jeth followed him to the barn.

"I got that coin cleaned better. It's from 1755."

"Wow. That goes with Evie's theory, doesn't it?"

"I think so." Jeth hesitated and set Pap's tools down on the workbench. "Did you know she was in that big pileup?"

"She was? Don't you kids tell me anything? She okay?" Pap's hands were back on his hips, his expression two parts concern and one part frustration.

"She's fine. She probably didn't want to worry you—"

"Nobody tells me nothin'," Pap grumbled and sauntered out of the barn, with Jeth trailing. Suddenly he turned back into the barn. "You're going to take me up to that cabin and show me where you found your treasures. Right now. I want to see it for myself, so I don't get in any trouble with the authorities for hiding an archeological site." He beelined for the four-wheeler parked in the equipment bay.

"Pap, the four-wheeler trail is blocked a good bit from the top. A tree fell a few years ago."

"So get the chainsaw. I can go for a little walk." He swung his leg over the seat and pressed the starter. The machine roared to life, puffing a little exhaust.

Jeth pursed his lips and checked that the chainsaw still had fuel, then placed it behind the four-wheeler seat. He sat behind Pap, their two bodies maxing out the weight the vehicle could handle. Pap zoomed away, flying around the edge of the fields like a teenager, not a seventy-seven-year-old man. He snaked the machine into the woods on an overgrown logging path, and they made it halfway up the mountain before a fallen tree too large to push out of their way or drive around stopped them. Jeth yanked the chainsaw to life, and in a few quick cuts,

Pap's access to the mountain was restored. Pap took the four-wheeler to where the road ended at the top of the mountain, then took off on foot through the woods, knowing exactly where the cabin was.

Pap was out of breath and sweating when they got there, but Jeth had to hand it to him, he was still pretty spry for his age. Evie had certainly whined more. She'd walked up from the bottom though.

Pap looked around the old cabin site. It wasn't hard to see the disturbed earth from a few days ago, even after the rainfall. He squatted down next to the hole and dug a little, but no new treasures emerged.

"Maybe it was an old medicine pouch someone buried, and the leather rotted away," he said, then straightened. He eyed the old cabin's ruin. "I don't think this site is any great historical landmark. At least not to anyone outside the family."

"Why would they build up here?" Jeth asked.

"Game trails for trapping. Maybe they timbered enough for building and firewood that they had a view. And years ago they had a water source with that stream. Heck, it still would make a great homestead, if you felt like building a road up here." He squinted into the woods.

"How old was the cabin, Pap?"

"Oh, my. It was a ruin when I was a kid. The walls were a little taller when my dad was a kid, but even then it had no roof. I liked hunting up here. Just sit a little behind that wall, on the hillside, and you can see down the hill into the state game lands. The deer pretend you can't see them. They sneak right along those posters."

"I always liked this spot, too."

"Hard to think this might be my last time up here," Pap said to himself. The comment caught Jeth off guard, and he didn't know how to respond.

In the silence, a stick cracked. They looked down the hill to see a doe trot through the brush, barely visible. Pap raised a single finger to point, and then she was out of sight, silently disappearing into the foliage. Another rustle sounded, and a buck appeared, his neck stretched out and nose working in hot pursuit. His white antlers shone in the bright yellow evening light like a crown. He paid them no mind, and then silently disappeared himself.

The two men strained to listen for the deer's return, but they were gone. They turned to each other and smiled, but something bothered Jeth. He toed the leaves with his sneaker. "You think that's a sign?"

"For what, kid?"

"That I'm supposed to buy the farm from you and keep it going?"

Emotions flashed across Pap's face. Shock. Happiness. Confusion. Sadness.

"Are you considering it, Jethro?"

He nodded.

Pap took a sharp breath. He removed his ballcap and ran a hand through his hair. "Nothing would make me happier. But your parents busted their butts to put you through college so you could have a good career. And you do. From what I hear, you're dang good at what you do. And I'm so proud of you." He hesitated, his eyes boring into Jeth as if to see his soul. Jeth kept his eyes down, unable to meet the scrutinous gaze. "Farming isn't as romantic as they make it seem on TV. Its long hours. Little pay. We work in all weather—"

"I know, Pap. I've watched you my whole life."

"And it's also the best way to live." Pap met his eye as Jeth looked up in surprise. "I mean it. I wouldn't trade a day. I still love beating the sun out of bed. I love the seasons, the changes of weather. I loved my cows—when we had them. I love working outside. It was a great way to raise a family. My girls are independent, hard workers, because that is how they grew up. You just need to know what you're getting into before you make a decision like that. It would be a very different life than what you've built."

"I know. I just hate to see it all gone."

Pap chewed his lip. "I think you should talk to some other farmers. I'm sure you've got some friends you went to high school with. You still hang out with Charlie?"

"Haven't seen him in a while. Would be nice to, though."

"Give him a call. Go help him out for a day. Then we'll talk some more." Pap laid a hand on his shoulder, and they started back down the path to the four-wheeler.

"Don't you have to give the developer an answer soon?"

"I'll make them wait. They waited two hundred years. What's a few more weeks?"

Chapter 11

After he and Jeth got off the mountain, Bill had a strong urge for beer and wings that night. He might be old, but he didn't have to act like it. He headed out to the local bar, missing his wife Grace with every mile he drove. She loved going out for dinner, like any good farmer's wife who tired of cooking would. Bill was pretty adept in the kitchen himself, but Grace had a way of making half-rotten tomatoes and squash taste gourmet. Oh, how he missed her.

He was pleased to see he wouldn't have to eat alone. His old friend Robert Smith was seated at the bar, beer in hand, watching the game on the big screen in front of them. His shiny bald head was unmistakable.

"Hey, Robert. How are you?" Bill pulled up a stool next to him.

Robert's face lit up. "Hey, Bill. It's been awhile. You keeping busy?"

Bill shook his old friend's hand. "Always." He ordered a Yuengling. "How's the family?"

"Good. Kids are all doing great. They'll come into town for Thanksgiving this year."

"Even Brian?"

"Yup, all the way from Colorado. His wife and kids are coming, too."

"That will be nice."

"It will. How's your family? The girls still working hard?"

Bill smiled and nodded. His daughters had been on overdrive their entire adult lives. Even now as they headed towards retirement, they showed no sign of slowing down. They'd be the kind of people who took on parttime jobs in retirement for fun. He couldn't say anything. He was past retirement and still working full time. "Grandkids, too. Evie's got her own bookshop in town."

"My wife loves that place. Seems Miss Evie has turned into quite the entrepreneur. She's made reading fun and social again. Town needs more of that fresh spirit." Robert took a sip of his drink. His tone turned somber. "How are you getting by—without Grace?"

"Miss her every day," Bill said. He was grateful that the bartender appeared with his drink in that moment. To lighten the mood, he slapped the bar. "This getting old nonsense is for the birds."

"Tell me about it!" Robert raised his glass, and they clinked glasses in a toast.

Bill took a long swallow, the cold beer running down his throat with the same timeless pleasure. His very first drink had been a Yuengling. If only he hadn't aged. If only Grace hadn't aged. But no one lived forever. Life was finite. He spun his glass on the cardboard coaster, studying the ring the condensation made.

"You miss your farm?" he asked Robert.

Robert sighed. "Yes and no. I love retirement. I don't miss the work. Financial security is something I'm still getting used to." He laughed. "And yet every time I drive by that warehouse, I feel like a traitor. I suppose it helps that the place doesn't look even remotely like it used to. It was never that flat. The trees and little hollows in the hills are gone." He huffed. "There was this one rock, a massive thing the size of a truck, that I had to plow around every year. It was always in the way. I broke equipment so many times when it came to harvest time and it was hidden under the crop. Well, they just came with their bulldozers, gave it a little push, and boom!" He clapped his hands together. "The rock was gone. They loaded it on the bed of a trailer and hauled it away. I fought that rock for so many years. My wife called me crazy, but I almost asked them to drop it off at our new house as a souvenir." He shook his head. "The ground is just metal and concrete now." He went silent, suddenly downing his drink in a few long swallows.

Bill stared at the TV that hung above the bar. A football game was being broadcast, but he didn't even register what teams played. "They offered to buy mine."

"Huh. How much?"

"Ten mill."

"Not bad."

"That what you got?"

"Don't forget they subtract taxes out of it. We pay taxes—on the property we own free and clear—for generation after generation. When we finally sell out, and the government reaches out their hand for more. I think I paid over a million in taxes on mine. But still, we were left with plenty." He thought for a moment. "You gonna sell?"

"I'm thinking about it."

"I can't advise you one way or the other. No next generation interested in taking over. The old bones givin' out. Only you can make that decision."

"Jethro said he's thinking about farming."

"Jeth? Since when? Isn't he one of those fancy engineers?"

Bill nodded. "Since today. I can't see him as a farmer. And yet I would *give* it all to him if he would be."

"Easy. You've got two other grandkids. And two daughters. You can't just give it all to one wanna-be farmer grandson that could change his mind as fast as he changes clothes."

Bill frowned. One part of his mind said *Why not?* It's how the farm had been passed for generations. The other acknowledged his other children and grandchildren would feel slighted. Getting nothing out of a ten-million-dollar inheritance was a slap in the face. He'd seen families divided over that kind of thing, sometimes the damage irreparable. He'd only had his brother Tommy, who was gone now just like Grace. Tommy had never married, never had children, which was probably why he'd stayed active with the farm. Bill had hired labor outside the family when he could afford it. The rest had fallen to his immediate family. But no, he didn't want his family squabbling over money. It was petty to divide family like that.

Robert ordered another drink. "It was nice to give my kids each a share. No fighting. They can do with the money what they will. It's how Brian bought his ranch out west. You know he's got a thousand acres? That'll be his family legacy moving forward. The tradition just moved west. And the other kids each got their share as well. I'm not sure Suzy is spending hers as wisely..." He frowned into his drink.

"I get what you're saying though. It gives them all an option. I suppose if Jeth really wanted to, he could buy another farm with the money." Bill was back to studying his coaster again.

"He could." Robert watched the game a while. Eventually, he could only shake his head with sadness. "I can't tell you one way or the other, Bill. You've got to weigh this one yourself."

Bill's wings came, Robert left, and still he lingered after his plate was empty. With the bustle and excitement of the bar around him, small as it was, he was loathe to go home to his big empty farmhouse.

Chapter 12

A few days later, Jeth was again asked to bale straw by Pap. In light of his consideration of being a farmer, he couldn't refuse. He got off work as early as he could, changed into old jeans and a t-shirt, and was soon throwing the bales into the loft. The challenge was not in the throwing, but in directing the handful of teenage boys Pap had talked into helping. They had the muscle, but they were still learning the technical side of stacking bales in a pattern that wouldn't topple and making sure they were tight enough to not slip a leg in between. Jeth had to move more than a few bales twice, which meant he was throwing around probably double what the teenagers were. At least it was only straw, and light. Just wait until these football players got their hands into the denser hay bales when the next cutting was ready.

Finally Pap pulled in with the last wagon, which the boys quickly unloaded. He put the baler away, then paid his help, except for Jeth, who wouldn't have taken payment anyway. Jeth dusted straw from his clothes and rinsed his face in the farmhouse sink once he and Pap were back in the house. He looked out over the fields as Pap cracked open an ice-cold beer for each of them.

"How would you ever do this by yourself?" Jeth asked.

"It's impossible," Pap shrugged. "That's why a lot of farms grow up in weeds. How did the boys do up in the mound?"

"They're teenagers. But they worked hard."

Pap smiled. "They'll learn. This sort of thing is the best way to develop work ethic. They're far ahead of their schoolmates who were home playing video games all summer." He tipped his beer to Jeth. "It sure taught you. And now look how successful you are. You did good up there today by the way."

"Thanks, Pap." Jeth mused at how much better beer tasted after hard work.

"I mean it, Jeth. Thank you for your help." Pap watched him with a scrutinous look. "Did you get to talk to Charlie yet?"

"Nope."

"Since you don't have a date tonight, why don't you swing by? I'm sure they're still out. This is their busy time of year."

Pap didn't push more than that, but Jeth could tell that today had been a test of sorts. Was he interested in being a farmer, as he claimed, or not? Maybe it *was* a good day to visit Charlie. He was already smelling like a farmer anyway. He finished his beer, thanked his grandfather, and then swung by Charlie's farm on his way home.

The big silos loomed tall above him, and the smell of cow manure filled his nose as soon as he stepped out of his truck. *Fresh Country Air* his mom always called it. And though the source was pretty gross, he had to admit it had a bit of a homey, sweet smell. He doubted any candle companies would be putting out Odeur de Cow anytime soon though.

He looked around for the Hess brothers, and eventually spotted Charlie up by the row of hutches, feeding calves. He walked up, carefully avoiding stray manure and mud puddles. Charlie's face lit up when he saw him.

"Hey, Jethro! Good to see you!" Charlie set down his pail of milk to shake Jeth's hand. "Long time no see. How's it going?"

"Good, Charlie. You?"

"Keepin' busy. What brings you here?"

"I was unloading straw at Pap's, and it put me in a farmin' mood. You want some help?"

Charlie smiled. "I won't turn that down, though I'm not sure how helpful you are." He nudged him in the ribs, but handed him a bucket of milk out of the back of a golf cart. "Here you go. One bucket per calf. Just set it in their bucket holder and go down the line."

Jeth couldn't help but smile as the three-week-old calves licked their lips and bucked in place in excitement for their evening meal. He barely set down the first bucket and the little calf had its nose buried in the milk, slurping greedily. He gave it a rub on its soft head. The calf behind it reached through the fencing and took a nibble on his shirt.

"Hey!" He pulled away, and quickly took a bucket from the laughing Charlie and fed that calf. He set the bucket in the pen, and the little cow happily drank. He stroked it's brown head. "Charlie, do you like your job?"

"Wouldn't want to do anything else."

"You work some insane hours though."

"Twelve to sixteen hours a day, six days a week. At least we worked it out among us that we each get a day off now. Dad didn't when he was our age."

"How does your wife feel about that?"

Charlie laughed. "Well, she insists that I'm home by eight for when she gets supper on the table. And she helps on the weekends. Every now and then we load up the kids and go away for a week. Again, I'm glad I have brothers, so I can have time off. But overall, I think she loves this place as much as I do."

"She's still teaching though, right?"

"Right. Thank goodness. I get to live off her benefits." Charlie puffed out his chest in jest, then laughed at himself, giving a calf a rub on the head as he started collecting the now-empty buckets. "We won't be rich anytime soon, if that's what you're asking. And don't ask me what I make per hour. Quite frankly, I don't ever want to calculate that out. Why are you asking, anyhow?"

"Pap got an offer for his farm." Jeth bit his lip. "I'm not sure I'm supposed to be telling people, so keep it quiet."

"Dang. I'm sorry Jeth." Charlie's whole demeanor changed, and he even stopped moving as he talked. "I hope he's selling to another farmer at least."

Jeth shook his head, kicking at a clod of mud. "Warehouse."

"But they can't build there. It's too hilly."

"You'd be amazed at what we can engineer these days."

"Wait, are you designing it?"

"Not directly. I don't even know what company made the offer. But I've worked on projects with worse looking land than Pap's."

Charlie sucked in a breath, then delved into his work again, collecting buckets with vigor. The happy cows still licked their lips, trying to keep their noses in the buckets until the very last moment. He turned back to Jeth as he stacked the buckets on the back of the cart.

"So you thinking of buying him out and starting to farm?"

"It's a crazy idea I guess."

"Not crazy. Just a big idea. It'd be a big change for you." He hopped in the golf cart, patting the seat next to him, which Jeth slid onto. "I mean, what do you make now, a hundred k a year?"

"One-fifty." Jeth avoided his eye.

Charlie let out a low whistle, then shrugged. "That's what somebody *should* make with a college degree in engineering." He pulled up to an old bank barn. "Got to feed the heifers." Jeth followed him into the open area of the lower barn, and watched as Charlie took a fifty-pound bale of hay in each hand and effortlessly carried them across the barn and into the central feeder. The cows crowded him, and he pushed them out of his way like he was one of them. "Mary, get moving. Jewel, get on your own side." Once he had escaped from the cows and thrown bales into two other feeders, he hopped into the cart again.

"You going to stay to milk?"

"Sure." Jeth glanced at his watch. It was seven o'clock. It didn't matter how late he stayed. It was worth the cost of this conversation. "Charlie, for real. Am I crazy to consider it?"

The golf cart pulled to a stop outside another building, this one far more modern than the bank barn, with steel siding and regular man-doors. The wear, rust, and dents indicated it was at least forty years old. Charlie led the way into the milk parlor, where his two brothers were already on the line cleaning cow udders and hooking them up to the milkers. The cows' hooves were at chest height, the men down in a little gully of the parlor so they had easy access. They smiled at Jeth as he followed Charlie in, but didn't pause in their work, wiping each udder clean with a paper towel and disinfectant before attaching the pumps, which then worked automatically until the cow's milk was depleted. Even as Jeth watched, fascinated, a cow finished up and the pump fell off and swung out of the way. The cow stood patiently as if nothing had happened.

Charlie handed Jeth a rubber apron with a smirk on his face. "I don't think you're crazy. I told you, I love this job. But it's going to be a life change for you. Not a career change. A life change." He handed Jeth a paper towel and a spray bottle. With only a minute of instruction, Jeth was cleaning udders from mud and manure so that they could be attached to the milker.

"So how's the engineering job, Jeth?" Charlie's oldest brother Kevin asked. He let one row of cows out of the stocks with the press of a button. When he pressed another button, the gates closed and a new batch of cows filed in, the stocks guiding them to stand with their back ends facing the milkers. One cow let out a large pile of manure, which splattered. Jeth stepped back, making a face. The brothers laughed, and the middle brother Dale stepped up with a hose to wash the worst of it away.

"It's fine. It's a job, I suppose."

"A good paying job," Charlie pointed out.

"Very." Jeth joined him, continuing to clean udders. He noted the other guys were cleaning two or three cows for every one he did, so he tried to move faster.

"Got a girlfriend?" Dale asked.

"Yup." Jeth smiled.

"Whoooo!" they hooted.

"Do we get a name?" Charlie asked.

"Sienna."

"Sounds like a country girl," Kevin said. He pressed the button to switch the cows on the other side of the milking parlor.

Jeth snorted a laugh. "I haven't seen that side of her yet. She's an engineer, too." Jeth bit his lip, smiling a little to himself.

"Even better, if she's making what you are." Charlie nudged him in the side, then reached around him to re-clean one of the cow's udders that Jeth apparently had not done satisfactory well. Jeth quickly re-wiped the rest of them, and attached the pump with a little sucking noise.

He thought he was getting the hang of things when Kevin cut in front of him as he reached for another cow. "Uh, not that one. She's a kicker. I'll get her." Jeth watched as the dairy farmer reached in, his arm carefully avoiding the cow's hooves as she stomped in protest. He cleaned the udders with quick efficiency, then got her hooked up. The cow settled down as the milk flowed into the lines in little squirts.

"How do you tell them apart?" Jeth asked in amazement. That cow's feet looked just the same as any of the other brown cows.

"Oh, you get to know them after a while," Kevin smiled. He already had the next cow prepped. Jeth hustled on to the next one.

Charlie cut in, "Keep in mind, we raise them from when they're calves. Like the ones you fed."

"So you breed all your own cows?"

The dairy boys looked at each other with eyebrows raised, never slowing their work.

"You do realize that's where milk comes from, right? Mama cows make it?" Kevin asked.

Jeth was still confused. "But what do the baby cows eat?"

"Oh, we still have plenty left over for them," Dale cut in.

"I'll let you try some fresh, warm milk when we're done," Charlie offered.

Jeth wasn't sure how he felt about that. He busied himself cleaning more udders. The work became automatic. The conversation helped. Sports, girls, their kids, politics—no topic was avoided in that milking parlor. Jeth realized that the three men knew each cow by name, often by the look of the udder. It was amazing.

He only got pooped on once, and learned how to use the hose to clean up the parlor.

When all cows had been accounted for, Jeth washed his hands with the others and took an obliging sip of fresh milk. It was warm, creamy, and he had to leave the decision until later if he liked it or not. He had the feeling warm milk was an acquired taste. Charlie clapped him on the back as he walked back to his truck in the dark of twilight.

"Hey, thanks for helping. We got done twenty minutes early thanks to you. My wife will be thrilled."

"All that and I only saved you twenty minutes?"

Charlie laughed. "You'll get faster with practice. You can come anytime you want. Hey, you want to come over for dinner? Laura always makes too much."

"No, I won't impose. I've got to get some stuff done at home anyway." The reality was that his mind was spinning, and he badly wanted to crash on his couch.

"Suit yourself. You're welcome anytime. I mean it." His face took on a more somber look in the light of the streetlamp. "Hey, for what it's worth, I think you should buy the farm."

Jeth was stunned. "Really?"

"A bicentennial family farm is getting rarer and rarer. I don't want to be the only one left. If your farm falls to the developers, it won't be the last. They'll come knocking on our door, and everyone else's. Your Pap has a beautiful property. You'll find a way to make it work. Doesn't have to be cows, either. Find what you love. There's good money in wine and corn right now. Or you could play with organics—"

"I can't pay what the developer is asking."

Charlie snorted. "Of course not. No one farmer can. Why do you think they offer that much? If your grandfather would sell it at farm-value rate, he'd have a dozen farmers knocking down his door. Myself included."

"Then what should I offer him for it?"

"Ask him what he needs to retire. He might surprise you." Charlie met his eye, then held out his hand. "Hey, good luck, Jeth."

Jeth shook his friend's hand and retreated, the smell of cows following him into his truck. As he drove home he thought about their conversation. He thought about the work, the affectionate baby cows, and the copious manure. He did the math in his head, which only spiraled into a headache.

He didn't know what to do.

Chapter 13

On Saturday afternoon, Sienna met Jeth at his house, then went with him across town to Pap's farm. Evie promised a good dinner as compensation for ruining his last few dates, though Jeth had warned Sienna several times about his reservations about Evie's cooking. He'd recited the tale of when she'd caught Grandma Grace's oven on fire making cookies, and they'd had to throw it off the back porch, a cloud of smoke streaming out of the house and across the yard. Sienna had blown off his warnings, but at least she had been warned.

Sienna's eyes went wide as she stepped out of her Jeep. "Wow. This place is gorgeous." It helped that her first visit to the farm was on a beautiful fall day. The air was crisp and clear. The leaves were turning into reds and golds. The light was working its magic across the horizon, though they were still a few hours from sunset. The oats were freshly combined, the straw baled, the tailings drifting in the breeze across the field.

"We love it." Jeth smiled and laced his fingers into hers. He really hoped she would like this place.

Sienna, for her part, had dressed with a certain enthusiasm. She had on a slim-fitting flannel shirt tied in a knot at her navel, skinny jeans, and the most fake, stereotypical cowgirl boots Jeth had ever seen. He'd tried hard not to laugh when he picked her up. She sure was cute though, and those jeans fit her hips just right, just like the country song on the radio sang.

"Come on, you have to meet my grandfather." Their hands swung between them as he guided her to the front door. He didn't bother knocking, just went in as always. Evie was in the kitchen, Grandma Grace's apron around her middle and a spatula in hand.

"Oh, yay! You brought her!" Evie jumped up and down with enthusiasm. She took off an oven mitt and held out her hand. "Sienna, it is lovely to see you. And since I ruined your date the other night, dinner," she twirled towards the stove and held out her hands, "is on me tonight." Her smile was real. And Jeth was grateful she didn't comment on Sienna's boots. Though when Sienna's back was turned, Evie mouthed, "What's with the outfit?", then giggled.

"Smells delicious," Sienna said and walked towards the stove. Evie had steak, potatoes, and sauteed veggies frying.

Jeth winced. "Oh no, I forgot to tell Evie you're vegan. Totally my fault." He was going on the bad boyfriend list for sure.

Sienna waved a hand at him, still studying the stovetop. "I'm not a true vegan. I just try to eat as clean as I can. I actually love steak. I allow myself to cheat every now and then."

Jeth felt a surge of relief rush through him. Another checkmark on Sienna's list of endearing qualities. "Where's Pap?" Better not let Evie see his infatuation, or he'd never hear the end of it.

"He's fixing one of the tractors. Again. Want to go tell him this is about ready?"

"Sure." Jeth reached for Sienna's hand, and she stepped into him, lacing her fingers with his. Hand in hand, they went back outside. It was nice to see Pap's parking lot full again, just like the days when Grandma Grace cooked Sunday dinner. Jeth noted the new car that could only belong to Evie and made a mental note to tease her about it. It was even brighter in color than the last one. They found Pap in the barn, up to his elbows in grease, his torso halfway inside the front end of Massey.

"Dinner's ready," Jeth said, peering over Pap's shoulder at the mess of an engine.

"Good. I quit anyway." He threw his wrench into his toolbox with considerable force, making Sienna start. "Old Massey is as old as I am. I told her she can't quit

until I do. Maybe she heard I'm getting too old for this."
He turned to face Jeth and Sienna, his eyes widening at
Sienna. "Oh. Hi there."

"Pap, this is my girlfriend Sienna. We work together at
Masters Engineering."

Sienna extended a manicured hand to Pap, "How do
you do, sir?"

Pap eyed her hand, hastily wiped the grease from his
own onto his dirty pants, then shook it firmly. A twinkle
lit his eyes. "No 'sirs' necessary here, Sienna. Just Bill will
do. Or Pap if you're so inclined. Welcome." He gestured
towards the house. "Shall we? Your cousin has been in
there half the afternoon. I'm not sure if that's a good or
bad sign."

Sienna smiled and followed them back into the house,
where Evie had put the last of the dinner on the table.

They sat, Pap said grace, and they dug in. Sienna was
two bites into her steak when she commented on how
good it was.

"Oh good! This was from Jeth's deer last year," Evie
said as she shoveled food into her mouth.

"Evie!" Jeth hissed.

The table went quiet. Pap cleared his throat. Evie
looked towards Jeth with a "What's wrong with you?"
expression. Jeth's eyes locked on Sienna, waiting for her
to react as she stared at her plate. She took another bite.

"I always heard venison is gamey. This isn't at all. Well done. How did you make it?" Sienna asked.

Jeth breathed a sigh of relief.

Evie blabbered on about her marinade until everyone was finished eating.

Stomach full, Jeth stretched out his legs under the table.

"So, Evie. We still have enough daylight to go hunt for gold again. Unless you discovered in your research that there is no gold in Pennsylvania."

"I didn't get to that part of my research. I got stuck in the history section. I was going to go back next week."

"Why don't you just call your old boyfriend?"

Pap gave his foot a nudge under the table. Jeth looked at him curiously.

"Tyler? No way." Evie took a long drink of iced tea and set the cup down with a punctuating thud.

"Never mind then," Jeth said slowly, watching Pap shake his head. "Just take another dive into the library. I'm sure there's something in there about rocks with quartz veins."

"Isn't that where they say the gold runs? With quartz? That's what they always say on TV," Sienna contributed, sipping her iced tea.

"We have a different kind of quartz here," Pap clarified.

"What kind?" Sienna asked.

"That's what I need to find out," Evie shrugged. "A lot of people say there's only coal in our mountains. It would be interesting if there was something else, too. I saw this story once, where some pioneer pulled a gold nugget out of the Susquehanna."

"That could have been dropped by someone." Jeth rolled his eyes. That story was probably an old wives' tale, and it was the only one people know.

"Who knows for sure? Glaciers moved a lot around." Evie's eyes drifted to the window, looking up at the mountain on the backside of the farm. "I wonder if a geologist ever took a core sample from that mountain. It's got something in it that attracts lightning."

"Lightning?" Sienna asked.

"Evie almost was struck as a kid," Jeth explained. "Now she's scared of thunder storms."

"I am not."

"Yes, you are." Pap laughed. "And I don't blame you. I'm scared of them too after watching you run home that day half-blinded."

"Wow," Sienna's mouth dropped. "That had to be terrifying. How old were you?"

"About ten," Evie said, sipping her tea as she glared at Jeth. He smiled back, thoroughly enjoying her agitation.

"It was wild. Her mother is the one that found her," Pap explained.

Evie shifted in her seat. Jeth raised one eyebrow, and let his mocking smile fade. Her demeanor had changed since she'd said Tyler's name, and it wasn't just because they were teasing her about her lightning fears. What had his old friend done to her other than break her heart last year? He supposed that was bad enough in itself.

"So Sienna, what do you do?" Pap asked.

"I work at the same engineering firm as Jeth. I'm in the environmental department."

"What does that entail?" Evie asked.

Sienna smiled. "Basically, I'm the one that surveys wetlands and designs erosion protection and drainage systems that will have the least environmental impact. I try to conserve topsoil, avoid excess water runoff. That kind of thing."

"Sounds a lot like what farmers do." Pap took a second helping of potatoes.

Sienna laughed, the sound filling the room like music. "It is, in many ways. At least the farmers that are doing right by their land, like you. I've found that a lot of you have already realized that soil and water conservation efforts are best for growing things. Some farms I've been to had stormwater systems in place decades before the government even started mandating them. It's fascinating how well you know your land."

Pap lifted his glass to her. "It's good that you appreciate that. Not all do. Jeth, you should keep this one. I like her already."

They all laughed.

When Jeth walked Sienna back to her Jeep that night, the stars were beginning to shine. She stopped and stared, taking it in. He gave her hand a squeeze. "You in a hurry to get home? Or I could show you something."

"It's a Saturday night. I have nowhere to be."

"Then come here." He took her hand and guided her up the worn path to the top of the little overlook, the place with the bench where Pap was rumored to have proposed to Grandma Grace. He pulled Sienna onto the seat next to him, and she laid her head on his shoulder, both of them watching the stars, the highway in the distance like a river of light on the other side of the fields.

"I like this place." She sighed.

He looked over his shoulder at her. The wind blew her blonde hair in wisps around her face, and the moonlight somehow gave it a silver glow. "I always pinned you as a city girl."

"I never got to be in the country like this. I don't know what I am, I suppose. Hey, you clean up pretty well yourself, for a country boy. I never would have guessed this is how you grew up. Or that you hunted."

"You don't mind, do you?"

"No."

"I am one generation removed, for the record. My mom grew up here. I just spent weekends throwing hay bales, picking apples, and roaming the woods. Whenever Grandma Grace would babysit, we'd pick blueberries or string beans...that sort of thing."

"That explains your exceptional work ethic. And that does still make you a country boy, by the way."

Jeth laughed.

A shooting star crossed the heavens, and Jeth turned to look at Sienna's contented face. She was stunningly beautiful. With a gentle, solitary finger, he turned her lips to his, tasting the sweet tea from dinner and a hunger that had nothing to do with food. Her lips followed his own, and suddenly he wondered if there was an energy in that spot that led men to do crazy things like propose. No, he wasn't ready yet, but for once he could easily see how it could happen. When their lips parted, Sienna let out a small sound of protest that sent a thrill to Jeth's toes.

"Your place or mine, tonight?" she whispered.

Chapter 14

Evie could hardly pay attention through the sermon Sunday morning. Her mind kept bouncing back to geology. Sure, she could spend hours that she didn't have in the library. Or she could simply call Tyler. Keep it professional. Keep it easy. See if he had an answer for her. Then move on.

Stupid Jeth, forcing her ex back into her mind. She'd almost had him out.

She played the conversation in her head. "Oh, hey, remember me? Well, can you tell me if there are any rocks in Pennsylvania that contain gold? What would they look like? Oh, there's not? Hey, don't laugh too hard." She gritted her teeth and tried to pay better attention. Ironically, the preacher was talking about focusing on God. How did preachers always seem to know when you weren't paying attention?

By the end of the service, she'd made up her mind to at least call Tyler. And she figured it was best to do it before she lost her nerve, so she called right from the church parking lot, her windows rolled up tight and doors locked, as if it was a secret conversation within the fish bowl of her car. Congregation members waved at her and she smiled and waved back, pressing the phone to her ear as her pulse pounded like a drum.

It rang twice.

"Evie?" a groggy voice answered on the other end of the line.

"Morning, Tyler." Oh lord, was she waking him? She glanced at the clock. It was ten thirty in the morning.

"You okay?" He yawned. Yup, she'd woken him.

"Uh, sure. Um, sorry for the timing, I just had a geology question for you." She heard a woman's voice murmur in the background. Oh no, it had to be *her*. "I should have just texted. I'll—I'll email you later. Bye!" She quickly hung up, her heart pounding and breaking all over again. She slapped her hand on the steering wheel several times. The car let out a loud honk, and a family passing her looked up in alarm. She grinned bashfully back, her face flushing with embarrassment, and turned the ignition. She was a fool to even try. That was why you were supposed to pay attention during church.

Evie spent the rest of the day cleaning the house from top to bottom. By evening she was seated at her kitchen table, her junk drawer dumped out as she sorted rubber bands, pens, matchbooks, and ill-placed screws. She still had not done more than a few quick Google searches into gold on the East Coast, and she had not drummed up the courage to email Tyler. He hadn't called back either. She hadn't expected him to. He was with his new girlfriend after all. Unless it was a new, new girlfriend. Who knew, with a guy like that.

She jumped as her cell rang and tensed as she saw the caller ID. He was calling back after all.

She held the phone in her hand, letting it ring three times before resolving to swipe to answer. "Hi," she said in a small voice.

"Hi. So what's your question?" His voice distracted her. It had a deep, musical, matter-of-fact quality to it that she'd always liked.

"Oh, uh—" *Keep it professional Evie.* "I've been trying to research gold—in Pennsylvania."

"We don't have much gold in Pennsylvania."

"Right. Just—wait...we *have* gold in Pennsylvania?" Her pulse pounded with excitement. "Where? What kind of rock would you find it in?"

She heard him take a deep breath. "Well, Pennsylvania gold is only found in trace amounts. Certain iron refineries can actually produce it as a byproduct, but they move a lot of ground to get anything substantial. The one in Cornwall is doing that. And there are some prospectors that like panning some of the rivers. They're more hobbyists. Really, that's it. What's this about, Evie?"

She didn't want to tell him about Pap's farm. He didn't deserve to know. "Well, how would the gold get here? Did they mine it like coal?"

He laughed at her, the sound like a thorn in her thumb. She reached for the rose anyway, waiting for him to answer. "No, it was deposited by glaciers from further north. Canada, or so they say."

She thought a moment. "So in theory, the glaciers could deposit large boulders, containing gold, on top of hills?"

"In theory. Some say that's how it got into the streams, washing downhill. What, are you taking up prospecting, Evie?" She could picture Tyler smiling on the other end of the line.

"Sort of."

"C'mon. What are you really up to?"

What could she do, lie? "We heard a legend about rocks being struck with lightning and finding gold inside. An old Indian legend."

"Where?"

"In an old book."

"You know that's not what I meant. Where are you looking? You panning gold in the Schuylkill?"

"You think there's gold in the Schuylkill?"

"Probably in trace amounts. I wouldn't want to sluice that much dirt to find out. You're avoiding my question though. What's the big secret?"

"The legend might indicate my grandfather's farm." She winced and closed her eyes as she said the words. It was too much.

Tyler paused for a second. She'd managed to surprise him. "Really? Did you find any?"

"No, I don't know what rocks to look for." *Oh, Evie, you're practically inviting him over.*

"Any quartz on the farm?"

"Lots of it."

"I'm happy to take a look. I don't think there's anything big, but you know I'm always interested in that kind of thing."

There it was. She bit her tongue and stayed silent, thinking hard. Half of their dates had been rock hunting, particularly for fossils. It had been fun. But they were dates. She was no longer dating Tyler.

"Look, strictly professional. You can even bring your boyfriend."

"I don't have a boyfriend," she blurted out. She cursed under her breath. "Look, I'm going to go. Thanks for the help."

"Hey, don't hang up on me again."

"Again?"

"This morning. Well, and a few months ago, too, I suppose—"

"That was a year ago."

"And I deserved it. Look, you have me curious now. Can I meet you tomorrow at your grandfather's? I always liked him. It would be good to see him again."

Evie gritted her teeth. Of course, he didn't mention liking *her*. Or wanting to see *her* again. "What about your girlfriend?"

"I don't have a girlfriend." He sounded surprised. Did he think she was that stupid or was the chick in bed with him at ten-thirty only a one-night stand?

"Look, I'll show you the rocks, and that's it. We just want to know what's on that mountain before Pap sells." She cursed again. "Look, I'm not supposed to mention that..."

"Secret's safe with me." He sounded like he was smiling.

"Okay, look, I get out of the shop at five. See you there at five thirty? Dress to hike, you have to climb a mountain." She was *not* going to sit on a four-wheeler with him even if it did mean a faster trip up the mountain.

"Sounds like a plan. Bye, Evie. I can't wait to see you."

He hung up, and she threw her phone across the table. What was she getting into? A whole year of tears, sewing her heart back together, and now here she was going hiking with the man? She pulled clumps of hair from her temples with both hands. The hairs held, but it hurt just enough to let her know she was alive. She hoped she still was tomorrow.

Oh Evie, you dummy. The library was a better idea.

Chapter 15

I t wasn't a good sign when the boss himself was waiting in the office at 7:47 on a Monday with the coffee pot fired up. Clint watched his staff with hawk eyes as they filled their cups quickly and filed into the conference room before taking off their coats. Sienna was seated inside already, a full cup in front of her. She didn't look happy, though she managed a weak smile at Jeth as he sat down across from her.

"What kind of fires are we putting out today?" he asked, stirring his coffee.

She leaned towards him, eyes on the door as their co-workers filed in, each looking more bewildered than the last. She whispered, "Rumor has it a RimForce rep spoke with Clint last week. They're changing the project."

"Which one?"

"Dunno."

She straightened as Clint came in and cleared his throat.

"Here's the deal, people. RimForce wants us to prioritize this next warehouse. They signed a contract with a global distributor, and they want it built yesterday. All of your other projects are on hold."

The room erupted in a cacophony of surprise and protest.

"I'm almost done my project for Doctor Warren."

"What, we don't need hospitals anymore?"

Clint raised and lowered his hands in a placating motion. "Look, all of our clients are important. Which is why I know you'll get the warehouse engineered quickly, so we can get back to our other work."

"Why even take on other clients if RimForce is the only one that matters?" The words were out of Jeth's mouth before he could think. His eyebrows went up in surprise at himself, but he didn't move to retract the statement. It was a legitimate question.

Clint glowered just the same.

Jeth rolled with it. "RimForce is our largest customer. Why not go exclusive? I'm sure they'll pass on more work."

Clint pursed his lips. Jeth knew full well that the man was the owner of the company and had pretty much married it. He'd given his whole life's work, two ex-wives, and his youth to building it into a multi-million-dollar behemoth with a reputation spotless enough to attract bigger companies, like RimForce. Jeth thought the big dogs treated Masters Engineering, Inc. like a servant. And though it was technically a hired hand, he wondered if that wasn't why Clint still held onto the clients they all actually enjoyed working for. After all, it was fun designing a CEO's private steel treehouse in the mountains. The architects got to have fun with creativity, but the engineers made sure it wouldn't roll down the mountain. And projects like the hospital one of Jeth's co-workers had in progress were rewarding in other ways. It all was way more interesting than squares of concrete and steel.

"Jeth, in my office. The rest of you, start calling your clients and tell them we have to delay a few weeks."

"*We* have to call them?" someone grumbled.

"Your phrasing will be, 'Our firm has taken on an emergency contract. We will resume your project as soon as possible. Your business is important to us.' Now get it done! I want overtime as much as you can."

Sienna snapped her notebook shut and leaned towards Jeth as she walked past. "Once again, acid mine drainage is pushed to the backburner." She looked down her nose at Clint as she exited. All but Jeth and Clint followed her.

"Might as well talk now," Jeth said, leaning back in his chair with hands folded behind his head.

"You have the brunt of this project Jeth. I need to know you're onboard."

"I'm always onboard. I just don't understand why we can't make RimForce wait like we do everyone else. Our little people's projects are important, too. Particularly to them."

"I know. Don't repeat this, but RimForce is paying double our standard rate to prioritize this. And as you know, our rate on this kind of project is already high."

"Why the rush? Is this the warehouse in Berks County? We rushed the prep on that one and then *they* were the ones that put it on hold."

Clint shook his head. "They're moving north, into vacant territory. They want it done so they're the first in the area up there. Sets the precedent for the rest of the area's development. The mom-and-pop restaurants and shops don't like their businesses next to warehouses. But warehouses like being next to other warehouses. If they can build an industrial footprint first, then they're more likely to buy up the other parcels and develop those to their liking as well. Kind of like how that big corporation did up in Hazelton. They've got a whole four-lane highway to their name now. I think that's where RimForce is headed."

Jeth furrowed his brow. "If they want to go north, why not just go up by Hazelton and build with the rest of the big developers? Or finish the project we started for them in Berks?"

"Aren't you listening? They want to be the first in the area. South Schuylkill is still a relatively blank slate. Isn't that where you're from?"

"Well, yeah."

"I thought that was maybe why you were pushing against the project."

"No, I didn't know that's what they were working on. My emails all were about the Berks project."

"New email. Sent it last night."

"Clint, why were you working on a Sunday night?"

"Double the price. So you're cool with the project?"

Jeth shrugged. "A warehouse is a warehouse. I'll get the design rolling once I have the specs."

"Good." Clint looked relieved, which made Jeth worry. He shook the man's hand as he left the conference room, and only when he sat down at his own desk did he put the pieces together.

Warehouse. South Schuylkill.

Pap's farm.

He suddenly felt nauseous. He opened his laptop, which seemed to load very slowly. He swiped open his work phone instead and opened his email app. There it was, within the long list of petty emails. They had a new project in southern Schuylkill County, not far off of Route 61. To be developed pending the finalization of the sale.

He bolted out of his chair and into the bathroom. The nausea won the battle over his nerves. There's nothing worse than hugging the porcelain of a public toilet. He flushed and leaned against the stall wall. It was happening.

Pap not only must be closer to selling, if RimForce was placing this project on such high priority, but Jeth was going to be the one to engineer the project that would bulldoze two hundred years of family history. How could he look Pap in the eye again? No amount of money could soothe that.

He was going to have to quit the project.

No. If he did that, then it would fall to some of the other engineers that surely would botch the job enough to ruin the farm completely. They would make sure every square inch was paved, the woods timbered, the mountain scraped away for shale fill. But if Jeth ran the project, with Sienna's environmental expertise, perhaps they would be able to convince RimForce that the warehouse had to be smaller. Or perhaps build two warehouses, and leave the stream running between them. Or keep them away from the base of the mountain, and set up a wildlife preserve on the hillside, claim it was too steep to bulldoze.

Nothing was too steep to bulldoze these days. That's what dynamite was for. They'd done it in past projects already.

Jeth had to run the project if he wanted any control of the design and development. He could ensure that the architects wouldn't make it an eyesore. He could do that.

His head was pounding, his brow and back wet with cold sweat. He'd face the project tomorrow. His stomach empty, he stopped at Clint's office on his way back to his own. He was sure his face looked green enough to be convincing.

"Hey, I think my cousin gave me food poisoning from her cooking. I'm going to have to try to work from home today. Promise I'll get done as much as I can."

Clint frowned but waved him out, already engrossed in his three monitors. "Feel better."

Jeth shoved his laptop and papers in his briefcase, and with a wave to Sienna, practically ran out of the office. He hoped to be at least partway down the road before the nausea overcame him again. After all, puking along the road *had* to be better than a public toilet.

Chapter 16

The broken tractor was taking up the better part of a day again, and Bill was almost at the point where he would be willing to call in a mechanic. While his arms were elbow-deep in metal parts, the sound of gravel crunching caught his attention. He looked over his shoulder to the house. Bill narrowed his eyes at the strange car, a new-looking Tesla SUV, that pulled to a stop. He wiped his greasy hands on a rag. A blue-suited man emerged and headed to his front door.

"Hello!" Bill yelled across the yard as he walked back to the house. "Can I help you?"

The man, a thirty-something, well-trimmed salesman-looking type of fellow spun on his heel to face him. His face lit up when he saw Bill. "You must be Mr. Fogel."

"Yes." Bill looked the man up and down. They were both on the sidewalk now. The man had the vibe of a car salesman, yet seemed genuine and enthusiastic. He was either new to the job or so good at it that he could appear likeable.

The man approached quicky, extending his hand. "My name's Todd. Todd Banks. I work for RimForce Development, Incorporated."

"Oh." Bill took the man's hand and shook it.

"I'm sorry to show up unannounced, but I've been calling for weeks now without answer. Perhaps we have the wrong number?" He pulled a paper from his back pocket, then rattled off Bill's phone number.

"No, that's right." Bill frowned.

"Oh. Well, good then. I apologize if it's a bad time. Perhaps you've been out of town."

"No. I've been here. As always."

The man's smile faltered for a second, then he carried on, sticking his hands in his suit pants pockets. "You see, Mr. Fogel—"

"Bill's fine."

"Bill...we haven't received your signed contract yet. My company is very interested in your property. I wanted to reach out in case you have any questions about the contract or the sale."

"Nope." Though Bill thought to himself, *Can you tell me where to get ten million without selling?*

"Excellent." Todd's smile grew brighter. "If you have it ready, I can take the signed form with me now."

"Didn't sign yet."

"I don't mind waiting a few minutes."

"No time today."

The smile did fade now, and Bill took a little inward satisfaction at that. Ah, the impatience of youth.

"Sir, I would like to be clear that the company will not uphold this offer indefinitely. If you decide not to sell, we will take our offer to the next potential seller. I'm sure you will find that we are offering well above the market price for this property. I, personally, would hate for you to miss out."

"Do I have a deadline? That contract didn't say anything about a deadline." The look in Todd's eye told Pap that all future contracts with RimForce would include one from now on.

"You're right, sir. Though it does state that time is of the essence. I will talk to the executives and see how much time they are willing to extend to you. Though I will warn you, they are highly motivated with this project. They may just decide to withdraw the offer. I can make no guarantees of how long it will stand." Dang, the man seemed to mean it, too.

Bill put his hands on his hips, meeting Todd in the eye. "Well, I'll be sure to give it another look when I'm finished my chores."

Todd's smile returned. "Excellent. Well, Mr. Fogel—Bill—it was nice to meet you. Here's my card if you do have any questions. And I'm happy to pick up the paperwork directly and save the post office the work. When you're ready."

Bill took the business card, his thumb instantly leaving a greasy smudge on the clean white cardstock.

"Good day, Bill." Todd shook his hand a final time, then headed to his Tesla. Bill smirked as Todd applied hand sanitizer within the confines of the car before driving away.

The gravel crunched as the car took off down the road, the faint whirring of the engine barely audible. Pap looked around the farm, realizing he really did need to make a decision. The corn was soon ready to cut. He was already prepping equipment to plant the rye, in part by habit, as a cover crop. But why buy or plant rye—or anything—if it was all going to be bulldozed in a few months?

He went into the house, the screen door slamming behind him with a smack-smack. He washed his hands in the sink, dried them on the dish towel, then settled into his wooden desk chair in the back room. The RimForce folder was on top of the desk. He cracked it open and scanned the first paragraph, with all its legal jargon defining the parties involved. Then he closed the file again. If RimForce was so eager, perhaps it was time to get a professional's help. He needed a lawyer that could translate the contracts and make sure that what Bill was reading really was what the words said, no loopholes included.

And maybe he'd hire a professional to help fix Miss Massey as well.

Chapter 17

Jeth's phone rang for the eighty-fifth time that day. Though his bowels had quieted down, he felt weak and exhausted. And he had not stopped working, at times even taking his laptop into the bathroom with him. He had a container of Clorox wipes at the ready. At least he'd managed to stay at his desk for the past hour.

"Hello?" he asked rather shortly, not even bothering to check caller ID. Or which phone he was answering.

"That's how I feel, too, if it's any consolation," Sienna's voice came through the phone.

He glanced at caller ID just to be sure. It was. "Sorry. Rough day."

"How you feeling?"

"Like I should crawl under the kitchen table. What about you? I told Clint Evie poisoned me. But you ate the same food."

"I feel great. It's just you. Though I suspect why. Have you seen the location of this new project?"

"Yep."

"Jeth, did you know? Why didn't you tell me? I mean, we were just there the other day."

"I didn't think Pap signed the papers yet. Maybe he did now. I thought I still had a chance—" He bit his tongue, realizing how ridiculous his words would sound in front of Sienna.

"A chance of what, Jeth?"

"A chance to buy the place," he spit out. He played with the pen on his desk, clicking it against a notepad, waiting for her to laugh.

She didn't.

"I wish you would," was her answer instead.

Jeth sat a little straighter. "Really?"

"I love that place. And it has so much family history. Why don't you buy it?"

"Sienna, you know what RimForce offered my grandfather? Ten million dollars."

"That doesn't seem like much."

His mouth dropped open, and he pulled the phone from his ear to stare at it a moment.

Sienna's voice continued on the other end, so he cautiously pressed it back to his ear. "I mean, honestly, they aren't making more land, right? And the stock market is going crazy right now. You never know if it's up or down. I wish my family had something we passed down generation after generation. We've lived in twelve states in three generations."

"Sienna, do environmental engineers get paid more than structural engineers?"

She laughed. "Banks have lots of money."

"But you need at least ten percent down...maybe more for a loan like that. That's my house, my retirement account, my car, and about the next five years of my salary just to come up with the down payment."

"I bet your ancestors paid everything they had to buy it, too. What *did* they pay?"

"Uh—Pap was telling Evie and I something about buying it from the Indians. Maybe a few hundred dollars? If that story is even true. It was probably gifted by William Penn's relatives or something."

"In those days a few hundred dollars *was* everything."

There was silence for a moment.

"You're serious, Sienna? You think I should buy it? I'd have to farm it then."

"Or lease it out to a farmer until you have it paid off from your day job."

"So you *are* serious."

"I'm on my way to see you now."

Jeth glanced at the clock and rubbed his forehead. It was almost six already. Past quitting time, at least for a normal day's work. "I kinda smell." And he was dressed in sweatpants. And the toilet probably should be cleaned.

"You have ten minutes to shower. I picked up dinner. Something easy on your tummy. Sweet potato soup with ginger." He could hear her smile through the phone.

"I better get moving then," he replied, then hung up and dashed to the bathroom. He felt just fine, and he would feel a whole lot better after a shower. And whatever weird vegan concoction Sienna had picked up did, at least, sound digestible.

As the hot water ran over his head, his mind spun. Where could he pull the money from? How much could he come up with? How much could he borrow? Would Pap even believe him when he told him he wanted to buy the place? He turned his face to the water, letting it run, then put his hands on the shower wall. He was an engineer. He crunched numbers for a living. Surely, he could figure out this equation as well. He just had to determine how many factors there were.

Chapter 18

Evie went into the farmhouse, looking for Pap. It was quiet, though the door was unlocked. She came back out on the porch, shading her eyes. His truck was still in the drive, so he had to be home. She spotted him far out across the field, towing the planter behind the tractor in neat, straight rows.

Tyler's truck pulled up alongside her new car. Her stomach twisted as if it felt the vibration of the big diesel engine.

"Hey," he beamed at her with a handsome grin. The truck door slammed behind him. With his hands in the pockets of his Carhart vest and blue jeans hugging his muscled legs, he didn't have the look of a scientist, though he'd constantly proven throughout their relationship that he was brilliant. He looked more like a model. Or the quarterback he'd been in high school.

"Hey," she tentatively smiled back. "Um, just let me leave Pap a note, and then I'll take you up to see the rocks." She ducked back inside the farmhouse. This was crazy. Insane really. Jeth could easily have met with Tyler, and she didn't have to get involved. They could have had guy time—bro time—whatever it was called. She scribbled a note to Pap on the tablet by the phone, then set it on the kitchen table. Emerging back onto the porch, she said, "Okay, a little hike if you're up for it. Unless you'd rather take the ATV?" She gritted her teeth at the idea of having to sit on the four-wheeler with him, holding onto his waist, legs touching.

"Hiking is fine," he said.

They set off across the field, Evie doing her best to lengthen her stride without appearing rushed. Tyler kept pace easily next to her. She wondered if he could feel the tension coursing through her veins.

"So how are you?" he asked.

"Fine. Great really. Super busy." She nodded with more vigor than necessary.

"Good. Glad to hear it."

"You?"

"Great. I work for the university now. College kids are" —he laughed— "interesting to say the least."

Evie had the sudden image of him bedding some college girl. Is that who had been with him when she'd called yesterday? She shook her head to clear the idea. No, even Tyler wouldn't mess around with his students like that.

"What's wrong?" He reached to touch her shoulder.

She stepped sideways. "Oh, just a fly, sorry." *Always apologizing, Evie.* "So, tell me more about gold in Pennsylvania."

"Well, we don't know too much about it. We know people find it if they look hard enough. We know glaciers scraped ore down from up in Canada, and there are small amounts all over the east coast. Sometimes nuggets. Primarily placer amounts."

"Placer mining is what they do on TV, right?"

"Yes. The dirt is washed, and then if gold is present they can sort it down to a few ounces per hundreds of tons. I think in Pennsylvania though, the amount is even less. Fun for hobbyists, not for miners."

"But you said there was a mining operation that was collecting it as a byproduct."

"There is. In Cornwall. But they make their money on the iron, not gold. It's a pretty cool process, really. They take the iron ore, refine it down like usual, then throw it into a microwave furnace, which separates out the gold and silver ore, which then can be refined. They actually use a magnet to draw out the powdered iron."

"That's fascinating. What kind of rock are they finding it in?"

"Weren't you listening? Iron ore."

Evie zipped her lips and nodded. She lengthened her stride a little bit. Better to get this over with. Then less of Tyler and less of this stinkin' mountain. She glanced at the sky. Were those dark clouds rolling in from the west? No, probably just her imagination.

The wind picked up.

Evie hesitated, now giving the clouds her full attention. Tyler passed her and started walking backwards in Jeth's fashion.

"We're going up, right?" he asked.

Evie nodded. The clouds were just clouds. She was being silly.

She followed Tyler now as he plunged ahead, as familiar with the mountain as Jeth after years of hunting with him. If Evie hadn't messed things up, he would have married into the family and been taken in as one of their own. Oh, wait. *She wasn't the one that cheated.*

They made it to the bottom of the hill, and started to climb up through the woods.

"I'm curious what kind of rocks are up here that you'd think there's gold in them. I hope they're worth the hike."

"I have no idea. It might be nothing. Hope you aren't disappointed. I did try to warn you."

"You sounded excited on the phone."

"I have been. There was an old legend, that on the Gobbleberg mountain, lightning would strike and the rocks would be cleft in two and there would be gold inside."

Tyler laughed, and somehow it didn't make Evie smile. It felt like he was laughing at her. "When is that legend from?"

"Eighteenth century. An old Indian told it."

Tyler laughed again, and Evie charged ahead up the mountain, beelining for the bottom of the rock ridge, the closest location where they could see the quartz-laden boulders. As if to match her mood, the wind was picking up with a warm breeze, and it seemed to be getting darker. She glanced at the sky through the trees. It was getting dark earlier, but it shouldn't be dark at six. Not yet. Surely, not a storm? She walked faster, the hair on the back of her neck prickling. *Get up the mountain. Get off the mountain. Don't be a wuss in front of Tyler.*

"Why don't we just run there?" Tyler teasingly complained behind her. When she glanced back, he had on a sly smile and wasn't even breathing hard. She ignored him.

A few minutes later they broke out into a small clearing, and the rocks burst upwards in a boulder pile like a miniature river of rocks before them. Evie gestured towards them with mock fanfare. Her attempt to be funny fell flat, and Tyler gave her a raised-eyebrow look like she was crazy. He plunged ahead, drawing a wire brush from a back pocket. He brushed lichen from one rock face, then leaned down to study it. Then he moved on and did the same to another. Finally he turned to Evie.

"So why'd you really bring me up here?"

Not what she was expecting. At all. "What?"

"C'mon. Even a layman like you knows sedimentary rock doesn't contain gold."

Evie felt the barb like a slap, as mild as it was. Her anger bubbled up. How was she supposed to know what sedimentary rock looked like? That's why she had called him. "You said placer gold is taken out of sediment."

"It doesn't work that way with sedimentary rock."

"Why not?"

"For one, even if there was gold, it would take forever to crush it and get the particles out. Particularly if they're in minute amounts like we know they are in this state. So in other words, you wasted my time. And I think you knew that. So why did you really lead me up here?"

Evie threw her hands in the air. "Oh, you know me, I just wanted to seduce you, Tyler. Did you ever consider that maybe I just have questions about rocks, and you could have taught me about them like I was one of your students instead of belittling me?" Her hands ended up on her hips as she glowered at him.

"So you were using me."

"Since when is asking your professional opinion *using* you? Ugh! You want me to pay you for your time?" Thunder rumbled in the far distance and suddenly Tyler didn't matter anymore. She turned and started heading down the mountain. Tyler followed a short distance behind.

"This is why we broke up, you know. You always want a favor."

"I requested your professional knowledge. You invited yourself over. And for the record, we broke up because *you cheated on me.*" She slipped on a log hidden by leaves and slid on her butt a foot down the mountain. "Ow!"

Tyler laughed.

Evie sat there, fighting back tears. The wind swirled the leaves around them. There was no doubt now, a storm was heading in.

Tyler reached down to offer her a hand. He still smirked, but Evie took his hand anyway. His warm, strong fingers closed over hers, sending an unbidden thrill through her. She let go as soon as she was on her feet. She wiped leaves from her pants—and Tyler from her hand. He plucked a leaf from her hair.

"I always liked your fire, Evie."

She didn't like the husky tone that had entered his voice. Nothing good came from that. Or perhaps because good things had come from that—that was the problem. Tears threatened once again.

"Tyler, I really did just want to learn about rocks. With the farm for sale, I just want to be sure—"

"There's no gold here, Evie. There never was."

Thunder rumbled in the distance.

She fidgeted, glancing from the sky to Tyler's face. "We've got to get off this mountain."

Tyler reached for her, like he was going to cup the side of her face to kiss her.

"What are you doing?" she asked, stepping back.

"C'mon, really? Isn't that why you brought me up here? I missed you, Evie."

"No, Tyler."

He pursed his lips, his expression changing to one of annoyance and disgust. He turned and started jogging down the mountain. She trotted behind him to keep up, but his long strides were unmatchable.

He paused just long enough to say, "You just can't get along with people, Evie."

She stumbled over a tree branch, anxiously looking again at what she could see of the sky. "You know, you're the only one who says that. I don't know how I'm supposed to 'get along' if you don't communicate! I was pretty clear about my expectations on the phone. Come and look at rocks. Answer questions about gold—"

"Stop yelling!" He spun on her, spittle flying, and she froze in her tracks. Before she could recover, he was another fifty yards ahead of her, barreling down the path.

The wind picked up, swirling around them. The leaves rustled with a low roar, a promise of more to come. Evie glanced at the sky in panic. It was rolling in. Fast. As it always seemed to on the mountain.

"Tyler!" she called, pushing ahead. He ignored her, his back fading through the trees down the mountain. That alone was enough to bring the tears. She watched him go until he had disappeared into the trees completely. A crack of thunder broke her willpower, and she crumbled to her knees, her head in her hands. The rain began in fat, heavy drops.

She felt alone. As the sky fell, her heart crumbled as it had been trying to do for a whole year, and the woods were all she had to hold her. She let the tears fall, grabbing earth in her hands, full of sticks and leaves that pricked her skin. The pain told her she was alive. That it wasn't imaginary. Her visions of how this day was going to go, *those* were imaginary. Tyler wasn't going to talk to her about rocks like an equal, like she was someone worth his time. And he wasn't going to come back to her with an apology. In his mind, nothing was wrong. Even though part of Evie recognized how horribly he treated her, the other part still longed for his approval.

It was so hard to fall out of love with him.

The rain came heavier now, drenching her. Her tears slowed and she turned her face to the clouds, letting the rain wash the pain away until she was numb. Let the thunder crack. Let the lightning flash. Bring it on. What more was there to lose?

The branches swayed in the wind overhead, and she became fascinated by them. The lightning cracked across the sky again, the bolts cutting veins of light above her head. Dimly, she noted that the hairs on her arms were standing up. She felt her head. Too wet to stand on end, it felt tingly, just like it had when she was ten years old.

The fear returned. The fight returned.

She pushed to her feet and started barreling down the mountain as fast as her feet would carry her, listening to the roar of the storm in her ears. She slipped on wet leaves, slid down the hill, regained her footing by clutching a sapling. There was a flash that seemed to envelop her, and the crack of the thunder was so close that she jumped and fell. She sat up and blinked. Her ears rang. The very earth around her seemed to vibrate. The smell of fresh cut wood tinged with smoke drifted around her as the wind and rain swirled.

It had missed. Again. But it had hit close.

She pushed to her feet. How many times could lightning miss?

Chapter 19

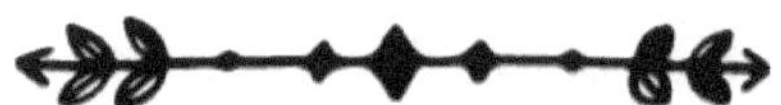

J eth and Sienna hunkered down in the old bank barn as the storm started. He'd brought her out here after they finished her delicious soup and had a long, productive talk. He was excited to talk to Pap about their plan to buy the farm. But Pap was still out across the field on his tractor, and it seemed awkward to just hang out in the house waiting for him. So he'd shown Sienna some of his old hideouts from his childhood, like the fort in the old chicken coop, and finished the tour in the old barn. Then the rain had begun.

At first they were content to stand inside the doorway and watch as the drops became larger and faster. When the deluge burst forth, Sienna stepped away from the mist that blew in from the open barn door. That was when she noticed the old, hand-hewn ladder.

Of course she was curious about the hay mound. Despite Jeth's protestations, she was surprisingly brave and nimble in climbing up the built-in ladder to the very top of the stacked pile. The sweet smell enveloped them, the green, soft second cutting a good cushion. Very quickly, Jeth realized she was curious about more than just stacking hay.

His protestations stopped. Their clothing became a nest, the rain and thunder the drums of their passion.

They were still in the mound, pulling their boots back on, when Pap pulled into the barn with the tractor and parked right below them.

"Oh crap..." Jeth whispered, and laid back in the hay, one arm over his forehead as his mind spun. Sienna giggled next to him, peering over the edge of the mound like a little girl. He shushed her, and she rolled back towards him, kissing him. He stifled a groan and pushed her off him, peering over the edge himself.

Pap got off the tractor, took off his hat to run a hand through his wet, grey hair, then sat on a bale next to the tractor, looking content to wait out the storm.

"This could take hours," Jeth whispered, crawling back from the edge to Sienna, who now had two hands pressed over her mouth to stifle her giggles.

Suddenly the ring of a cell phone cut through the noise of the pounding rain. Sienna's eyes went wide, and she pointed at him. Jeth flinched and grabbed at his phone, swiping the call to mute. In the process he saw the caller ID. Pap. He made a face at Sienna, showing her the caller ID. As he did so, it rang again. Sienna reached out and swiped answer, then waved the phone at Jeth to speak.

"Uh, hi, Pap. What's up?"

"Well, I was going to ask why your truck is in my driveway, but you may as well come down here and tell me yourself."

Jeth rolled to the edge of the hay mound and peered over. Pap was peering straight up at him. He waved once. Jeth smiled bashfully and waved back, phone still pressed to his ear.

"You coming down?" Pap asked into the phone.

"Yup." Jeth quickly rolled back from the edge and hung up. Sienna burst out laughing, the sound surely carrying down to below. Jeth glared at her. "I love how you find this funny."

"I feel like a teenager again." She pecked him on the cheek and climbed over him to the ladder, beginning her descent.

Jeth could feel himself turning five shades of red as his feet finally found the floor of the barn. Sienna was standing next to Pap, watching the rain. Pap shot Jeth a sideways, knowing glance.

"The rain on the tin roof makes a great sound, doesn't it?" Pap asked.

No one had to answer. Sienna beamed, shooting Jeth a wink. He wanted to melt into the floor, his face burning as if singed by lightning.

They listened in silence to the roar of the storm outside.

Eventually Pap nodded towards the house. "Any idea where Evie is? And that your truck, Sienna?"

Sienna shook her head. "I don't have a truck. My Jeep is at Jeth's."

Jeth frowned. "We didn't see Evie anywhere. I gave Sienna a bit of a tour of the place while we were waiting for you to finish up."

"Hmm." Pap raised an eyebrow.

A man jogged through the rain towards the truck. Jeth recognized him instantly and stepped to the edge of the open barn door, cupped his hands to his mouth and yelled, "Yo, Tyler! You guys okay?"

He waved a hand towards them. "Fine!" he yelled back through the rain. He opened his truck door.

"Where's Evie?" Jeth shouted.

"She's on her way down." Tyler climbed in his truck, and it roared to life. He pulled out of the drive without another word.

"On her way down from where?" Pap asked, his brow wrinkled with concern.

Jeth chewed his lip as he thought, concern rising within him. "They must have gone up to look at the rocks on the mountain. Sheesh, I didn't think she'd actually call him. Much less go for a hike with him."

Pap put his hands on his hips. "Jeth, you should know your cousin better than that by now. She's still head over heels over the boy. And quite frankly, I hope this is the end of it. I never much liked the man. All those years of letting him hunt here with you, and he never once said thank you." He eyed the torrents of rain. "Evie would never go up there in a storm. Didn't she see it coming?"

"You didn't," Sienna said quietly. She was right. Pap was partially soaked.

"I was trying to finish the row," he explained. "It's gonna be hard to get back in that field to finish after this deluge."

They all watched across the field towards the mountain expectantly, but Evie didn't show.

"I'm sure she's almost back," Jeth said and checked his watch. He knew Tyler and Evie had had their disagreements, but leaving her on the mountain during a storm was a low blow.

They waited five minutes, the downpour unrelenting, before Pap suddenly stood. He pulled a raincoat and rain pants from a hook and pulled them on. In a minute, he looked like a mid-Atlantic fisherman in his bright yellow suit.

"Pap...what are you doing?"

"My little girl is out there. Someone's got to go find her."

"I'll go, Pap," Jeth offered.

A crack of thunder resounded across the property. There was no doubt the bolt had struck somewhere nearby.

Pap walked over to the four-wheeler and fired it up, ignoring Jeth with a posture that told him it would be moot to argue. Their family had a stubborn gene. "I'll be back soon. If the rain slows down go head to the house and get some coffee brewing. I think we're all going to want it."

Chapter 20

S he didn't hesitate as she came out of the woods, charging ahead into the fury of the storm across the newly cut field. The rain whipped at her face. Her shoes squeaked from the water inside. Her jeans grew heavy and began to stretch.

She ran faster.

A four-wheeler appeared up ahead, headlight bouncing as it traveled fast down the old dirt road. Her legs pumped. The storm flashed overhead, a flash too bright, too close. She slid in the mud, barely catching herself. Then the four-wheeler was there, slamming to a stop. The engine rumbled with a comforting familiarity.

Pap reached a hand out to her. "Oh, honey."

She threw herself into his arms, sobbing into his raincoat. He held her tight a moment, then patted the seat behind him. "Come on, let's get out of this storm."

They rode back together, Evie's head tucked into her grandfather's broad shoulder, her arms twined around his waist like she'd done when she was a little girl. The storm died down to a steady drizzle as they walked into the house. Evie didn't know what to do in the doorway. She stood dripping on the mat as Pap pulled off his yellow rain gear. He ran a hand briskly through his hair, droplets flying. His shirt was damp at the neck and shoulders.

Jeth appeared from the hallway with two bath towels, handing one to each of them. Pap threw his around Evie's shivering shoulders, then pulled her into a firm hug. She sagged against him, still holding her folded towel to her chest like a teddy bear.

"When Tyler came down without you, I knew there must have been a fight. You'd never be up there in a storm on your own," Pap said softly into her hair. "You alright?" He pushed her away and checked her over. Jeth quietly stepped away, into the kitchen.

"I'm fine. But—that relationship is over. Pap, why doesn't anyone want to be with me?"

He laughed. "Oh honey. There's a lot of us that want to be with you. You're a vivid, inspiring personality. You throw yourself into everything you do. It intimidates some people, that think they're living their best life until they compare themselves to your enthusiasm. And they feel they either need to ignore or crush you in order to control you. You just have to hold out for that one person who is as alive as you are, and sees you for who you are. In the meantime, hang out with the rest of us as much as you can. Cause we like havin' you around." He smiled into her eyes, a hand on each shoulder.

Weakly she smiled back. It was hard to argue with that face. That didn't mean her mood felt much lighter.

She heard Jeth's voice coming from the kitchen. "What's Jeth doing here?" she asked, sniffling.

"You know, I haven't gotten the chance to ask him that yet. He brought his girlfriend back, too."

Evie glanced around, looking for Sienna.

"Go upstairs and find some dry clothes. I'm sure your mom has something hidden away up there still. Then come down and we'll have some coffee."

"Thanks, Pap."

A few minutes later, and much dryer, she padded barefoot into the kitchen in a turtleneck and jeans with a waist so high it was practically up to her chest. Pap was at the window, looking out over the fields, cup of coffee in hand. He'd changed into a dry shirt, but his jeans were still damp. Jeth and Sienna leaned against the counter next to the stove.

"Hey," Evie said to Sienna, forcing a polite smile on her face despite her own emotional exhaustion.

Sienna smiled and nodded back.

Evie went up to the sink next to Pap. The view from the window was stunning. With everything freshly washed, the light did something magical as the sun cut under the cloud layer. Every raindrop sparkled, even those on the mud of the corn field. A pair of geese flew over the pond, their honks audible even in the house. The land was alive, and they were nestled in a warm, dry place within it.

Jeth and Sienna already had coffee cups clutched in their hands. Evie smiled gratefully as Jeth nodded to a steaming cup that sat next to the coffee pot. She scooped it into her hands, grateful for the warmth.

Smack-smack went the door.

"Well, isn't this the unexpected party?" Evie's mom entered. "Isn't it a little late for coffee?" She crossed the kitchen and kissed Pap on the cheek. "Hi, Dad."

"Hi, honey," he smiled down at her.

"Why are you wet?"

"I had to rescue Evie."

Her mother turned to her with a furrowed brow.

"I'm fine, Mom."

"Are those my old clothes?" She pointed. A smile brightened on her lips, and she plucked at the fabric with nostalgic amusement.

Evie winced.

Her mother laughed and pulled her into a hug. "You better keep them. It's not like I'm ever going to fit into them again. So, what am I missing?" Her mom pulled off her raincoat and hung it by the front door.

Everyone exchanged glances, their faces mirroring the emotions battling in their minds. Pap leaned back against the sink and took a long sip of coffee, then cleared his throat.

"Bess, you better pour yourself a cup of coffee," Pap said.

"I'll never sleep—" Evie's mom said.

Pap motioned to the coffee pot with his thumb. "We're burning daylight here. I've got chores to finish, but we have things to talk about. Anyone that can't sleep is welcome to stay and help me."

Bess sighed and pulled a mug from the cupboard. She filled it halfway.

"Now, let's start with Evie. Why on earth were you up on that hill during a storm?" Pap narrowed his eyes at her.

"Evie, you didn't—not again!" Bess was silenced as Pap held up a finger to her.

Evie took a deep breath. "Well, Jeth said on Saturday that we should get a geologist to look at the rocks. So I called Tyler. And he wanted to look. So we went for a walk. And then—we had a fight."

"And he left you up there?" Jeth asked quietly.

Evie nodded, replaying her pathetic sobs in her head. "Lightning sorta does strike twice," she muttered.

Pap's eyes grew wide. "That strike...was that near you? Again?"

Evie nodded. "Too close for comfort."

Her mom pulled her into her arms. "Oh, honey. I'm glad you're okay. And I thought you and Tyler broke up?"

"We did, Mom."

"But then—"

"I told you, we need a geologist!" Tears threatened at the corners of Evie's eyes again. She hid in her mother's shoulder to hide her trembling.

The room was quiet.

Jeth finally cut the silence, clearing his throat. "I'm sorry, Evie. I didn't think when I said to call him."

Evie leaned into her mother's warm embrace and let her kiss her head.

"Did he look at the rocks?" Pap asked.

Evie shook her head, refusing to look up. All the dramatics with Tyler had led to nothing. No serious information that she couldn't have found online. Eventually she looked at Jeth and Sienna from over her mother's shoulder. "So what about you two?"

Sienna cleared her throat and looked away. Jeth glanced from Evie to Pap and quickly back to Evie again. "We just came to talk to Pap," he said.

"About?" Pap prompted.

"So much is going on. We can talk about it later."

"Jethro Fogel-Mueller, I'm not getting any younger. Spill." Pap set down his coffee mug and folded his arms. The old man was pretty intimidating, Evie thought.

Jeth swallowed and straightened. "You know how I was asking you about buying the farm?"

Pap nodded once, then hid his expression as he drank from his mug, which he'd snatched back into his hand as if it was a shield.

"I—and Sienna—would like to make you an offer."

Everyone's eyebrows went up but Sienna's. She laced her fingers with Jeth's and smiled at him.

"Well, we did some math and together, we think we can afford it." Jeth met Pap's eyes, his fingers tight around Sienna's.

Evie's mom cut in, "Oh, are you two getting married? And you didn't tell me?" She pulled away from Evie's clutch slightly as she looked to Sienna's hand for a ring.

Jeth cleared his throat, breaking his gaze with Pap to look at the ground. "Well, not exactly. Not yet."

"Oh. Well how would that work then?" Bess asked.

Jeth turned red. "We'll buy the farm together. If things work out, maybe we'll get married down the line, but if not, then we sell. That puts us in the same position it does now. Right?"

"Not exactly," Evie cut in. "Pap needs the money for retirement, and you can't possibly pay what the warehouse people can pay." She scrunched her forehead. "Can you?"

Pap's solemn voice cut in. "You're serious about farming?" All eyes turned to him.

Jeth met his gaze. "I'd like to. Eventually. We'd lease it out until after we've paid a reasonable part of the mortgage, just to keep the land cleared. And yes—" He looked at Evie. "I do intend to pay the developer price."

"It's not worth that," Pap said dismissively, waving a hand in the air. "A farmer would never see a return on that."

"Which is why I'd have to lease it out while I work my current job—"

"To who?" Pap narrowed his gaze at Jeth, and Evie was grateful she wasn't under that scrutiny. "All our neighbors are either selling or cutting back or letting their properties be overtaken by weeds as they go work other jobs. Or they're so busy with what they have, maxing out their land and bodies, that they can't handle more." He pursed his lips in thought. "Did you talk to Charlie?"

Jeth nodded.

"Then you know how hard it is to turn a profit farming, right?"

"But I'll have my engineering job. I can do the farm in the evenings."

"And never see your wife and family?" Pap shook his head. "Jeth...I know this is hard for you. You're attached to this place. But it's only a place. Farming is a whole *lifestyle*."

"One that we're interested in pursuing." Every head in the room turned to Sienna, who had spoken. "He wants to try, Mr. Fogel. And with what I have saved, and two engineering salaries combined, it's not impossible. It at least buys us more time to decide if we want to farm. If they build a warehouse on this land, we definitely never will."

Evie studied the couple carefully. They hadn't been dating long. Sienna was a total city girl. And yet they got along well. It was nice to see her supporting Jeth, while also being a total boss in her own right. Sienna was as smart as could be. Evie could see how it could work. And yet, if it didn't work, it had the potential to devastate the family further. Evie couldn't see getting Pap's hopes up for a next generation farmer, only to watch them shatter if Jeth quit after a few years.

"Honey, I think you need to think about this some more," Bess crossed the room to lay a hand on Jeth's shoulder. "Talk to your parents about it."

"I had to talk about it tonight. RimForce isn't going to wait around patiently. Pap, you have to know you have another offer on the table."

The earnestness in his eyes pinged Evie's heart, and she started looking from stoic Pap to Jeth's bubbling emotion.

"I'll think about it," Pap finally agreed.

"Mr. Fogel, you don't have long. RimForce is pushing the project through. In their minds, this is a done deal." Sienna met his eye.

"I just spoke with the salesman today. He said I have time." Pap set his empty mug on the counter with a clink.

Sienna inhaled, then pushed on in a rushed breath. "They're going to force a sale quickly if you don't agree to their terms. And the terms will not be as good if you make them wait."

Pap turned away and leaned against the sink, facing the window. "That's about what the salesman said."

"It's your right to take your time. Do *not* let them pressure you. But do be prepared to walk away or lose their offer if you don't make a decision soon," Sienna continued.

"Are you a lawyer?" Bess asked, her head tilted to one side.

"I need one," Pap grumbled, speaking into the sink.

They all turned to face him with varying degrees of surprise.

"You all need to read that contract. There's a whole lot of pages for something that should be a simple pay-this-and-sign kinda deal. I just want to be sure I'm not missing anything."

"My dad is an attorney in Philly," Sienna said and shrugged. "He specializes in real estate."

Bess's eyes lit up. "Perfect!"

"Mr. Fogel, please consider Jeth's offer as well. I was telling him—my family never had anything to pass down. What you have here is special. Quite frankly, I don't think you can put a price on it, but I also understand how doing so is necessary. Only you can weigh all the options."

Pap nodded.

Evie downed the last of her coffee, her mind buzzing. It sure would help Jeth to be able to afford the farm if there was gold on it. Pity such a thing was impossible.

"Dad, what else do you need to get done tonight? The sun's almost down." Bess looked at the wall clock.

"Oh, I—uh, it's going to be too muddy to finish planting. But I could use help filling the planter back up with rye, so I don't have to do it tomorrow." He looked hopefully at Jeth, who smiled.

"Guess I'll go practice being a farmer." He took Sienna's hand and with a mock salute to the family, slipped out the door with her in tow.

Bess kissed her father on the cheek. "I'm heading home, Dad. Let me know if you need any groceries tomorrow. I'm going to the store."

"I need some milk."

"Got it." She pulled Evie into a big hug. "Listen, honey, if you ever need to talk..."

"I know, Mom. I'll be fine." Bess planted a big kiss on her forehead and headed out the door, leaving Evie and Pap alone.

Pap leaned back against the sink, rubbing his forehead.

"Pap, you can't sell." Evie's voice was quiet.

"I can't live forever, either, honey."

She turned away and poured herself a second cup of coffee. "Why does God take all that is good?"

"Oh, he doesn't take it all. Sometimes humans do that." Pap seemed as if he spoke to himself. He suddenly turned back to her.

"You hungry?"

She shook her head. The coffee was about all she could stomach right now.

A twinkle rose in Pap's eye. "Want to play a round of cards?"

"What game?"

"Whatever you want to play."

They played a round of Go Fish until the sun set. Evie glanced at the clock, her coffee gone, even the caffeine not fully reviving her.

Pap seemed to read her mind. "You better get home, Evie." He stacked the cards and slipped them into their box.

"What are you going to do, Pap? Say you do sell, to Jeth or the warehouse people...then what?"

"I find a realtor with a nice cabin on a lake somewhere. And I fish to my heart's content."

Evie smiled. "Somewhere close, I hope. Where I can visit?"

"Of course."

Evie chuckled. "Pap, before you sign anything, we better get you a realtor. There aren't any lakes nearby."

Chapter 21

The realtor's office was just in town, and the woman behind the desk had a smile as bright and alert as Grace's had been. She wasn't much younger than Bill, with her grey hair pulled back in a youthful ponytail. She rose as he came in and extended her hand.

"I'm glad you could come, Mr. Fogel," she said brightly.

They shook hands. Hers was lotiony soft; Grace had had callouses to rival his own. Pap gritted his teeth to focus.

"Now what can I help you with today? What are you in the market for?" she asked, oblivious to his inner thoughts.

Bill settled into the chair next to her, a large computer screen and a book of flyers in plastic sleeves in front of them. He wondered if he shouldn't have worn a clean change of clothes, instead of coming in directly from the farm. Hopefully his deodorant was still working. Old Spice hadn't let him down in over sixty years.

"Ms. Wiley, I'm looking for a small place to retire. Ideally on a lake somewhere."

"Are you a fisherman, Mr. Fogel?" She smiled. She had very good teeth. Maybe a little too much lipstick.

"Yes, ma'am."

"Well, then, your best bet is someplace up on Lake Wallenpaupack. It's up towards the Poconos. Prime location. Five thousand seven hundred acres of water. Some of the best bass fishing around here. And we happen to have several lake houses for sale. What's your budget?"

"I'm not sure yet..." Bill furrowed his brow. "Don't you have anything closer than the Poconos? Awful lot of touristy people up there."

Ms. Wiley scrolled through her listings, then hit a few keys. "Well, there's always Lake Wynonah. That's about as close as you can get. Now that's a gated community, which frankly, you might love. Quiet, nice lake..."

"I'm not sure I want to be in a gated community." How would his family drop by unannounced? Besides, he didn't like rules telling him what to do with his land. "If there's nothing closer, what about something along a river? I like fishin' for trout, too."

Ms. Wiley's face lit up. "I've got just the place then. But it is a little bit higher priced. Gorgeous house on fifty acres. I suppose that's still downsizing, for a farmer like you." She laughed.

The realtor made a final click, and a gorgeous house filled the computer screen. She leaned back in her chair, hands folded, gauging his reaction.

Bill didn't know what to think. He leaned towards the screen. The house was stunning, with board and baton siding in a dark brown, a metal roof, and big windows. But what really got him was the manicured landscaping. There wasn't an inch of mud or twig out of place. Whoever was selling this place had really wanted to take care of it. Even the trees in the big lawn were trimmed. He reached for the mouse and clicked ahead. Every exterior shot was immaculate, though already he suspected the house was double the size of the one he was in now. Then there were more pictures of the grounds, with a lawn that stretched all the way to the Schuylkill river, a graded gravel driveway, and four-wheeler paths in the woods. There were two expensive, heated tree stand huts built into the forest.

He pointed at one, curious if it was part of the sale.

Ms. Wiley read his mind. "Included. I take it you're a hunter, too?"

Bill nodded. "Used to be. Can't walk too far anymore." A nice hut like that might make hunting season enjoyable again. He clicked into the interior of the house, which was, as he suspected, huge. Four bedrooms, three bathrooms, a kitchen with shiny stone countertops and white cabinets. Who picked white for cabinets anyway?

The living room centered around a big-screen TV. Bill thought it odd that the couches didn't face the big, floor to ceiling windows that overlooked the river. It was beautiful, but the more he clicked, the more he had a hard time picturing himself in the place. It seemed like a place for Jeth or Evie.

"It's lovely."

"It's even better in person. We can go look at it now if you'd like. The owners have already moved. The furniture remaining is only staged, though they're willing to sell the furniture separately if the buyer is interested."

Bill clicked back to the view of the river. It was pretty, but it was distant from the house.

"What do they want for it?"

"It's a steal. One point five million."

His eyebrows rose. He didn't need a house that expensive. Not unless it came with more than fifty acres. "Anything else?"

"I have a little place along the river in Port Clinton. But that's a little more of a fixer-upper. And it's in the flood plain, if that matters to you."

Bill looked at the little ramshackle cottage that appeared next. It was cute and just the right size, but he didn't need to deal with a fixer-upper. Or floods. He sighed. Were the real estate pickings really this slim?

The realtor smiled. "It can be overwhelming. Look, now that I know what you're in the market for, I will keep an eye out. New listings pop up all the time. Just be ready to act fast, before they're snatched up."

Bill nodded and rose, Ms. Wiley rising with him.

"Thank you for your time, ma'am."

"Anytime, Mr. Fogel." She shook his hand, and though she still smiled, the spark was dimmer than when he'd entered the room.

She didn't smile like Grace after all.

Bill's phone rang as he slid into the front seat of the truck. He flipped it open and saw the strange number but answered it anyway.

"Hello?"

"Mr. Fogel!" A warm, booming male voice said from the other end. "My daughter apparently is quite a fan of yours. She's been pestering me all morning to give you a call."

"Who?"

"I'm sorry, sir. My name is Mark Vonn. I'm an attorney in Philadelphia. I believe my daughter Sienna is dating your grandson, Jeth."

Realization hit Bill. "Oh, yes. Sienna is a wonderful young lady. We've only met a few times, but she seems like the real deal."

"She's a good kid. Listen, she said you've had a developer contact you. If you need someone to look over the contract, I can. Pro-gratis, on Sienna's order."

"That would be wonderful. I can pay you though, Mr. Vonn." Bill was never one for handouts, and yet a lawyer with a recommendation from someone he trusted—like Sienna—was necessary.

The lawyer laughed. "If you heard my rates I'm not sure you'd want to. No, don't worry about it. This is the first my daughter's asked me to do anything like this for someone, so how can I refuse? I'm sure it will only take a few minutes to look over. Those kind of contracts are usually pretty straightforward."

"I just need a second set of eyes to tell me it's saying what I think it's saying."

"Certainly. Can you fax it over? Email?"

"Uh..."

Vonn chuckled. "How about I have Sienna pick it up and scan it? I can look it over tomorrow afternoon then?"

"That sounds like a plan." Bill felt his face turn an indignant shade of red. These young people, even the old young people, were so tech savvy. Yet ask them to change the oil in their own cars—

"I'll be in touch after I look it over tomorrow, then. Have a good day, Mr. Fogel."

"Bill is" —the call cut out— "fine." Bill looked at his phone and frowned.

He started the engine to the truck, reveling in the rumble that vibrated through the seat. Enough wasting time. He had work to do.

Chapter 22

The call with their ten-year client had not gone well. The man on the other end was a builder redesigning and expanding a hundred-year-old theatre in a busy downtown area. They were trying to follow all the local building codes, thus why they'd hired an engineer, but now an entire work crew was sitting on their hands because Jeth didn't have the project done. He had RimForce's new warehouse pulled up on two separate computer screens. They were likely going to lose the job with the builder, even though it was halfway completed. It was bad business all around, but Jeth wasn't the business end of the firm's machine. He was just an engineer. And right now, he felt his reputation was on the line. It would be his stamp on the drawings, his name.

He pushed off from his desk and walked over to his boss's office, knocking on the door. Clint also worked in a fishbowl, and Jeth could practically see the smoke coming out of his ears through the glass. Clint had his phone pressed to his ear as he typed in a steady staccato. He waved Jeth in.

"I know, I'm working on it. Let me talk to the guy I have in charge of that segment, and we'll get it done. I promise. Yup. Bye." He dropped the phone to the desk without his hands ever leaving the keyboard. Jeth mused over what kind of case the phone must be in to survive that kind of abuse. "Now what, Jeth?"

"The builder on that old theatre project is mad. Can I get that job done so they can roll with it? I had it halfway before this latest RimForce assignment."

"How long will it take you?"

"A day or two. I have to visit the site to get some measurements."

"No. RimForce wants this by next week. We can't spare the time."

"He's threatening to find another firm."

"And until they get him in, he'll lose even more time. He's just bluffing. And he will wait like the rest of them." He continued typing, then pounded the enter key and turned his full attention to Jeth. "I asked if you could handle this project. I hope you told me the truth. If you're struggling..."

"I'm fine. But we have other clients that we promised work to."

"They aren't paying double."

"You know that's not fair."

"Life isn't fair."

"Oh please. Seriously? You built this dang company on these little people you're now slighting!"

"That's right. I built this company. So why do you care what I do? I hired you to design. Not advise me. Just do your job!" He turned back to his computer, clicked through a few things, then began typing again.

Jeth left the office without another word. His fingers flexed as fury bubbled within. He had a pretty steady temper, but not with that kind of treatment. He glanced at Sienna as he slowly made his way back to his office. She was on the phone, but met his eye with an eyebrow raised. He pursed his lips and slipped back into his own office, then sat staring at his computer screens for a good ten minutes, accomplishing nothing.

Sienna came in.

"Hey, you okay?" she asked.

"Nope. But it is what it is. This project sucks."

"We'll get it done. It's not like we're the builders."

"I like it better when we work for people that are trying to rebuild and improve the community. Not bulldoze and pave it."

"Jeth, you design this stuff. You're out of a job if you don't get it done." She shot a nervous look towards Clint's office. "Look, I've got to keep moving, too. But I wanted to let you know I talked to my dad. He's going to call your grandfather and look over the contract."

"Pap can't afford that."

"I know. Which is why I told him to do it pro gratis."

"Seriously, Sienna? You know how much of an insult that will be to Pap?"

She looked taken aback. "I was just trying to help…"

"You're so desperate for my family farm that you'll even get your dad involved. He's who you're getting the money from, isn't it? Why doesn't he just buy it?" Sienna looked as if he'd slapped her. Jeth saw the bite of his words and instantly regretted it. "I'm sorry…"

To his surprise, she stayed calm and merely shook her head. "This isn't you, Jeth. This project is getting to you. You need to get out and clear your head tonight."

"But we were going to meet up with your friends…"

"I'll go alone. I'd rather go alone, than be around this side of you. Go find the Jeth I fell in love with, the one that wants to follow in his grandfather's footsteps and make his community better. Then we'll talk." She turned on her heel and headed back to her own office.

Jeth turned back to his screens and began to calculate. Slowly, the design of half a million square foot warehouse came into the picture. It was hard to concentrate as Sienna's words replayed in his head.

He spun in his chair to look at Sienna through their twin fish tanks, tapping his pen on his desk. She was focused on her work, phone to her ear as she typed, her blonde hair slicked back in a professional ponytail.

Had she really just said she'd fallen in love with him?

Pap was in the barn that evening, cursing at Massey. The old tractor had quit on him again. Jeth peered over his shoulder.

"Going well?" he asked sarcastically.

"I'd be just fine if I could get a new hydraulic pump. But they stopped making them twenty years ago."

"I hate to say it, Pap, but maybe it's time to send Massey to the graveyard."

"What? Never!" His tone was shocked, but there was a twinkle in his eye. He looked Jeth up and down. "You going up the mountain?"

"Yup." Jeth was in full camo. With a few hours of daylight left and perfect fall hunting weather, a few hours in the woods would do his mind well.

"Good luck, kid." Pap smiled and turned back to the tractor.

After crossing the field and entering the trees, Jeth snuck through the woods with the ease of a practiced hunter. He only had a few hours before dark, prime time for the deer to come out. It was the perfect day for it. Crisp but not cold. Perfect sunlight with only a few puffy clouds in the sky, and minimal breeze.

He carefully scanned the ground before placing each foot, careful to avoid the small sticks that would crack and alert any game in the area. The leaves rustled lightly, but they were still colored, and thus not as dry as they would be in rifle season, when every step would crunch. A stick cracked under his foot, and he stilled, listening, making sure his prey had heard nothing. Then he slowly traveled on, climbing the mountain a footstep at a time. A light sweat broke out in the small of his back, and he was grateful that he'd left his heavy coat down in the truck. He got to the base of the rockslide and stepped onto a boulder. It was harder work to walk on the boulders up this part of the trail, but quieter. He balanced carefully, his compound bow in one hand, as he worked his way up. The birds chirped in the trees, but otherwise the woods were silent.

The smell of fresh cut wood was heavy in the air, and grew stronger as he climbed. Halfway up the boulder field, he froze. Before him stood a massive pine, split down the middle with a black streak. Below the tree a boulder had split, shards chipping from the sides where lightning had headed into the earth. He glanced around. There was no doubt, the damage was fresh. Some shards were even on top of the newly fallen leaves. He knelt down.

Part of the rock had been turned to glass, black, shiny, and yet still crumbly with pockets of gravel. It was like dirty obsidian. Inside the rock was a curious shade of yellow that reminded Jeth of gold. There were only flecks of it, and it could just be some kind of iron stain. Surely, it must be some other mineral.

It couldn't be gold. There was no gold in this part of the country. Evie had asked Tyler, showed Tyler, and he had concluded it was too rare, that it was only found in placer amounts. This Jeth could see with his naked eye.

Leaves rustled to Jeth's left, and he stilled, senses pricking. There was a monster buck in the woods this year, according to the trail cameras. And big bucks were smart bucks. He couldn't make quick movements or noise without being detected. He nocked an arrow, his pulse quickening. His eyes scanned back and forth as much as they could without him having to turn his head. He breathed to steady the nerves that thrilled through his body.

A squirrel jumped through the leaves from the spot he'd heard the rustle. Jeth gritted his teeth. Squirrels were the noisiest creatures in the entire forest. A bear or a deer, hundreds of times their size, could walk as silent as a ghost, while a tiny little squirrel made the noise of an elephant.

Jeth waited a moment longer, then un-nocked the arrow, restoring it to the quiver. Hefting the bow in his hand, he slipped a sliver of the shattered boulder into his pocket, the flecks of yellow evident in just two little spots. Then he continued on, stalking as quietly as his boots would allow, hoping to make it to his tree stand before the monster buck headed out into Pap's cornfield for the night.

Chapter 23

Bill rubbed his chest where the nagging ache was worsening. He'd felt a little funny this morning and taken an aspirin. It had made things feel better. He'd gotten what was left of the planting done, but now the pain was growing again. Grace would have told him to go get it checked out. Perhaps he should. But Grace wasn't there anymore. He popped a second aspirin and downed it with a full glass of water. Leaning his hands on the farmhouse sink, he looked out over his fields. This getting old nonsense really was for the birds. No matter how much he cut back, no matter how much help he tried to hire, no one could do the work the way he wanted it done. So he kept going. He enjoyed it, too. It made him feel young.

Or at least younger.

He'd tried leasing with some success. The neighbor farmer had been appreciative. But the rent barely covered the taxes on the place. Bill mused on what the taxes might be for the warehouse people. Surely quadruple the cost of what they were now. But the builders wouldn't even blink at a cost like that. What did warehouses rent for per month anyhow? Probably more than what he paid in taxes for an entire year. He'd heard a rumor that some of the big companies negotiated with the townships to make sure the tax rate would stay low. Their argument seemed legitimate. More jobs for the area. More people to tax, more income to tax, more services rendered. It was good for the local economy.

Bill rubbed his chest again, the ache continuing.

Old Massey was now as fixed as she was going to be, with a band-aid repair job that would last a few more weeks. Maybe they would retire together. He was supposed to be spraying the fruit trees at dark, when the wind died down. There weren't many left, but the kids liked the apples. Even though his two little girls were all grown up, he still pressed a batch or two of cider for them every fall. He needed a lot of apples just for one batch. He looked to the sky. It wasn't long before dark. Maybe he could get the apples sprayed, and then if his chest was still bothering him, then he could head into urgent care, just to be safe.

He clapped his hat onto his head and headed back out. It took until he hooked up the sprayer to the tractor, filled it with water, and mixed the chemicals before he had to sit down, the pain increasing to where he felt a bit queasy. He took off his hat and was surprised to feel the sweat beaded across his forehead. Grace would have yelled at him. He had to do it. He had to get checked out. For her sake. She'd made him promise not to rush off to Heaven to meet her. She said she needed a break for a few years, let him take care of the kids for a while, and then she'd have a nice place all ready for him when he got there. He could still picture her laughing as she said those words, the countenance of a joke and yet ringing with all sincerity. She did that a lot. He understood. It was her saying she loved him, and to fight on as long as he could, until the good Lord called him home to join her.

Hopefully that day wasn't going to be today.

He noted Jeth's truck in the drive, wondering how the hunting was going up on the hill. Should he call him? Naw, why interrupt a perfect hunting day for an old man's indigestion. He retrieved his wallet from the kitchen counter in the house. The screen door slapped shut behind him, but then on impulse Bill pulled it open again to pull the main door shut. He didn't lock it. Jeth might want something in the house. Then he pulled himself up into the truck, wiped sweat from his brow as he caught his breath, and then drove himself to the local ER, fifteen minutes away.

The receptionist took one look at him and ushered him into a back room, where he sat on the bed, the ache in his chest worsening. His mind whirred through the possibilities. Indigestion. Heartburn. Simple fatigue.

Maybe that was it. He hadn't slept well the night before, not after all the excitement with Evie and Jeth and Bess and Sienna. Then the realtor this morning. Maybe he did need to slow down a little. Go fishing. He didn't need a cabin to do that, just a rod and his truck and a parking spot along the Schuylkill River.

A doctor appeared a few minutes later, stethoscope in hand.

"Hey, Bill, how you doing?" the young Doc Antony asked. His father and Bill had gone to school together back in the day and had stayed in touch somewhat over the years. Of course, Doctor Antony Senior was gone for a few years already. So many of his friends were gone: a side effect of old age. He hated to think of how things would look when he was ninety.

Doc Antony pressed his stethoscope to Bill's chest.

"I'd be better if I wasn't getting old," Bill grumbled.

"I hear you're having chest pain." The doctor frowned as he listened to Bill's heart.

"A little. Probably just being dramatic."

A long silence stretched while Doc Antony listened to the stethoscope, periodically moving it to a different place. When he straightened and removed the earpieces, he looked concerned. Wordlessly he put a blood pressure cuff on him, watching the dial as the cuff tightened. When the pressure released, he looked at the EKG.

"It's a good thing you came in this time, Bill. You're having a mild heart attack."

"What? Naw, I'm fine. Just a little indigestion."

Doc shook his head. "I suspect you have unstable angina. I can't be sure how bad it is until we do more tests and bloodwork, which we'll get going in a minute. I'm going to order you some nitro, which should stabilize things for now. But I'm going to have to admit you for the night. We need to keep an eye on you."

Bill was silent. The news made him more nervous than the pain. "That serious, doc?"

"That serious, Bill. Honestly, if you'd kept working tonight instead of coming in, you may have gone into a full cardiac arrest. And being on that farm alone—that could have been it."

Bill felt a tremor of emotion run through him. His hands started to shake.

"I don't tell you that to scare you, Bill." Doc Antony laid a hand on Bill's shoulder. "I just know how you farmers are. You need to understand the seriousness of the situation. Now, you *are* going to be fine. And once we know more about what's going on in there, after the tests, we'll have a picture of what we need to do going forward so you've got plenty more years to work. Okay?"

Bill nodded, not trusting his voice.

"Now, I'm going to send in a nurse with the medicine. Then once they locate a room for you we'll transfer you upstairs for the night. You want to call one of your girls?” Doc Antony raised an eyebrow. “Did you even tell anyone you were coming in?”

Bill chuckled half-heartedly. “I didn’t want to scare them. Better call Bess though. She doesn’t live far away.”

Doc squeezed his shoulder again. He checked Bill’s chart, where Bill had listed Bess as his emergency contact. “Sounds like a plan. You’re in good hands, Bill. We’ve got you.”

As Doc Antony walked out, Bill felt a ball settle into the pit of his stomach. *Being on that farm alone—that could have been it.*

Jeth came down from the hill in the dark, guided by the moonlight, his eyes adjusted to see in the dark. He loved hiking in the dark. There was something magical about being out in the crisp moonlit air, with stars sparkling overhead. It made him feel one with nature. He often stayed in his tree stand until the sun was completely set, well past viable time to shoot, just to listen to the forest come alive around him.

Unfortunately, tonight he was still empty handed. He loaded his bow into his truck and shut the door, intent on checking in on Pap. Only, it didn’t appear Pap was home. The house was completely dark, not even a porch light on. The streetlamp by the barn glowed bright, but that was on a night/day sensor. And Pap's truck was gone.

Jeth crawled in the front seat and retrieved his phone. There was a voicemail from Evie. He wanted to tell her about the lightning-blasted rock anyway, so he pressed call without listening to the message. The phone rang only once.

"Jeth! Where are you?" Evie's voice sounded panicked.

"At Pap's. Why?"

"You're at Pap's?" Her tone turned to confusion.

"Yes..."

"Then why didn't you drive him to the hospital?"

"Wait—what? I was up hunting on the hill. What do you mean drive him to the hospital?"

"Mom just called me in a panic and said Pap drove himself to the hospital *while* he was having a heart attack. She's headed there now."

"He did *what?* He was fine earlier. I saw him before I went out."

"Well he's not now! He's having a heart attack!"

Jeth's own heart hammered in his chest. "How bad?"

"I don't know!"

"Evie, calm down." He said the words as much to himself as to her. "Can we see him?"

"I guess, since Mom's going in."

"Okay, I'll lock up here, and then I'll pick you up along the way. Okay?"

"Okay," she said in a small voice. "Jeth, I'm worried. We just lost Grandma Grace…"

"He'll be fine. The docs can do whatever he needs in there." He wished he sounded more convincing. "See you soon." He ended the call.

He jogged over to the farm house and scanned the kitchen to make sure there wasn't food on the stove or anything crazy. There wasn't. Nothing looked like Pap had rushed out. He rummaged around the key rack next to the door until he found the key that fit the lock to the front door. It took some effort to even find the key, but finally he locked it. He hid the key under the doormat, not knowing who the next visitor to the house would be. Or if a spare key existed. Who knew when that front door had even been locked last?

Then he headed out to pick up Evie, his knuckles white as he gripped the steering wheel, a million thoughts running through his head.

"Hey," Evie said as she slid in the front seat. Her eyes were red, visible even in the glow of the car lights.

"You hear anything more from your mom?"

"No." She sniffed and leaned with her elbow against the window as Jeth pulled out.

"What hospital is he in?"

"The new one." She snorted. "At least he didn't have to drive far. I can't believe he drove himself in. Mom's probably giving him an earful."

Jeth laughed, and it relieved some of his stress. Aunt Bess was certainly one to give you her opinion on a good day, to say nothing of when she was stressed.

"How was the hunting?" Evie asked.

"Beautiful. Didn't see anything but squirrels though. Oh, and I found where you almost got struck by lightning. Again." He glanced behind him into the back seat. "Hey, reach in my jacket pocket."

Evie reached behind them. "There's aren't any boogie tissues in there are there?"

"Nope. Just a present for you."

Her brow furrowed as she dug in the jacket pocket, withdrawing a small rock. She tossed the jacket back into the back seat and flicked on the light over the passenger seat. Jeth waited for her reaction, biting his lip in excitement.

"It almost looks like…"

"Gold."

She looked up at him. "But Tyler said that's impossible."

Jeth shrugged. "You said he didn't look at the rocks. Maybe it's pyrite. We've got that in Pennsylvania. But whatever it is, it *looks* like gold. So your old legend is true, at least in what the old Indian that told it perceived to be gold."

Evie turned the rock over in her hand. "Huh."

"You've gotta come up the mountain and check out the boulder. It's pretty cool."

"No way. I'm never going up that mountain again. I don't want the third time to be the charm...and I actually get struck."

"The chances of that are like—"

"Me and that mountain combined attract lightning!" she protested.

Jeth pulled into the Emergency parking lot. He sighed as he turned off the truck. "One thing at a time. Let's go see how Pap's doing."

Evie led the way into the hospital, and they were directed at reception to a room upstairs on the third floor.

"I always hated the smell of a hospital," Jeth whispered in the elevator.

"They don't smell like anything. It's the sanitizers," Evie murmured back. The elevator dinged, and they stepped out into a well-lit corridor. A sign overhead said Cardiac Wing.

"They do have a smell. Do you have to argue everything?" Jeth wrinkled his nose.

Evie glared at him. "Oh, then what does it smell like?"

He didn't answer.

They stopped outside Pap's door. Jeth knocked once, then entered the open door. They could see Pap on the far side of the room, Aunt Bess at his side with his hand in hers.

"Hey, kids," Pap said in a weak but cheerful voice.

Evie fell into his arms, her cheeks wet with tears. "We were so worried."

He patted her back. "I'm fine. I'm fine. Doc says after the tests I should be able to go home tomorrow. This is all just a precaution." He waved at the heart monitor and other electrodes attached to him under his hospital gown. The monitor beeped at a steady rate, the squiggly line drawing itself across the screen. "I feel better already. They gave me some medicine, and that seems to have done the trick. I've got a bit of a headache, but the chest pain stopped." He looked at Jeth, smiling. "How was the hunting?"

"Just squirrels," Jeth shook his head and smiled, leaning against the windowsill next to Aunt Bess.

"I just gave him an earful about driving while having a heart attack," Aunt Bess smiled wearily. "Particularly since you were there."

Jeth shook his head. "I didn't even have my phone on me. Left it in the truck so I wouldn't spook the deer. I guess from now on I just have to mute it. And check it."

"Oh, Bess, I was fine. All this is so mild," Pap protested.

"You are not *fine*, Dad! You're *in the hospital.* Thank God it wasn't worse."

Pap frowned and leaned back against his pillows. "So, Evie, how's the treasure hunt going?"

"What treasure hunt?" Aunt Bess cut in.

"Jeth found this up on the hill today." Evie handed Pap the rock.

Pap turned the rock over in his hand, then looked at Jeth. "It looks like glass."

"Lightning did strike up there. It blasted a boulder no less, and a tree above it. There were pieces everywhere. It looks like it melted the rock."

Pap squinted at it closer then shook his head. "I wish I brought my glasses. Pretty neat find."

Evie pointed at the flecks in the rock. "Pap, there's little pieces of yellow in it. We don't know if it's pyrite or gold. It's pretty though."

Pap smiled. "Maybe you found the treasure of the Gobbleberg after all."

"Gold?" Bess interjected again. "Evie, is that why you and Tyler were up on the mountain last night?"

Evie nodded. Understanding dawned over Aunt Bess's features, but before she could ask more questions, a man came in, a male orderly with a stern face. "Hello, Mr. Fogel. We are ready to take you for your chest x-ray now." He started organizing wires, preparing to move him. The three family members stepped to the side.

"How long will it be?" Aunt Bess asked.

"Oh, about a half hour or so. You're welcome to wait here. Or the cafeteria."

Jeth waited for a joke about hospital cafeteria food, but it never came. Within minutes, Pap was being wheeled out.

The three of them looked at each other across the empty room. Aunt Bess rubbed her forehead and sighed.

"They told me they weren't going to run the tests until morning. I wonder if they found something that concerns them?" Her face was lined with worry.

"I'm sure they just want to run them while the machine is available." Evie hugged her mom, and Jeth watched awkwardly as his tough aunt took a shaky breath.

When mother and daughter parted, Aunt Bess's strength had returned. "Well, I'm going to hang around until he's done. May be a long night. You kids should head home. I'll call you when we learn something. He seems fine. It's a miracle he had the wherewithal to actually come in."

Jeth gave her a hug. "Should I call my mom?"

Aunt Bess gave him a squeeze as she pulled away, wiping a tear. "I did already. She knows. I'll keep her in the loop as well."

"Is she coming home?"

Aunt Bess shook her head. "No need. Pap sounds like he's going to be fine. And she's due to come up for

Thanksgiving anyway." She sighed. "I do wish she was closer at times like this, though." She squeezed Jeth's hand. "Your mom said she was going to call your sister, too. Honey, if you need anything, you let me know, okay?"

"Well, same for you, Aunt Bess."

She cupped his cheek. "Thanks, Jeth. Okay, you kids go. Enjoy your night. I'm going to go sample the cafeteria food. This is a new hospital. How bad can it be?"

Aunt Bess rode the elevator down with them, and Jeth told her every horror story involving cafeteria food he could think of. She was swatting his shoulder by the time the elevator landed, threatening to give him coal for Christmas, but they all were smiling. As Jeth and Evie left the hospital the mood turned back to somber. They got in the truck, shut the doors behind them, and sat in silence.

"What if it was worse?" Evie whispered. "It could still get worse. He could need surgery."

"One day at a time, Evie. That's what Grandma Grace would say. That's all we can do."

Chapter 24

B ill watched the weather the next morning with a glum face. Beautiful and sunny, highs in the seventies, and he was going to be stuck in bed. He was exhausted from a late night of being poked and prodded. Wasn't it ironic that people came to a hospital because they were sick and needed rest, and yet the hospital was the hardest place to find rest? People moved around the clock. Alarms went off down the hall at odd hours of the night. And the lights—would it kill them to turn off a few lights at night?

Bill changed the channels, flicking through them without interest. Is this what normal people did in the morning? He had work to do. He'd missed perfect spraying weather last night for the apples. Perhaps he should have finished them before coming in.

It could have been the end. The doctor's words still echoed in his ears.

Where was the doctor, anyway? Doc Antony had stayed down in the ER and assigned him to Doctor

Patilovskyikov—or something. Bill was embarrassed at his inability to say the name on the cardiologist's nametag, despite asking him to repeat it several times. No wonder he wasn't on the doc's priority list.

Bill settled back on the weather channel again, and sipped the orange juice out of its plastic cup that was left over from breakfast. They'd told him to start cutting back on the coffee and bacon. As if that was going to happen. It was going to rain mid-week next week. He was going to need help to get the soy beans cut. That would take some long hours on the tractor. The good Dr. P had said tractors were going to be off limits for at least a week. Bill didn't understand why. It's not like it was strenuous to let a machine do the work. Imagine back when his grandfather was a kid, and they did it all by hand.

He picked up his flip phone and dialed Chuck, fellow farmer and father of Jeth's friend Charlie.

"Hey, Bill!" Chuck answered brightly. "How are ya?"

"Could be better, Chucky. I'm in the hospital."

Chuck's tone fell instantly. "What? You okay?"

"Mild heart attack they tell me. I'm fine though. But I'm gonna need some help getting the soybeans in while the weather's good. I know you and the boys are busy yourselves, but I'm happy to pay ya. Heck, with the way things are going, you could probably just take the crop and put it in your own barn. I hate to see it go to waste."

"Sheez, Bill. Yeah, we can help you out. A heart attack?"

"So they tell me."

"Well, hey, you take care of yourself. We'll take care of the crop. When are they letting you go home?"

"Hopefully today," Bill glanced at the clock. It was nine-thirty. The day was half over.

"Well, if you need anything else, just call."

"Thanks, Chuck. Talk to you soon." They hung up.

A few more minutes of clicking through commercials later, and a doctor came in.

This doctor must be on the new shift, as she was bright faced and smiling. And he could pronounce her name, Greene.

"Good morning, Mr. Fogel!" Dr. Greene said brightly.

"Morning. I hope you have good news, Doc."

"I do. We have a pretty good picture of what's going on with your heart, and right now it looks like you won't need surgery. She rolled up a computer screen and clicked through to the image of his x-ray. Pointing with the tip of her pen, she said, "Your chest is clear. The EKG results from this morning are much better. I think with some blood pressure medication we can make things a bit easier for your heart so you don't have further issues."

"I never had issues with my blood pressure before..."

"That's good to hear. When was the last time you had it checked?"

Bill thought back. When was the last time he'd seen a doctor? Had it been that long?

Doctor Greene smiled. "Well, we caught it now, and we have medication that can keep it stabilized. We see this a lot in aging patients. Now, I want you to take it easy for a few days. If any symptoms return, you need to come in right away. And I want to see you in a month for a follow-up. Okay?"

Bill nodded.

"Great. I'll see you next month then. You can pick up your prescription and set up the appointment at the desk. Enjoy your day, Mr. Fogel."

"You, too. And thanks." He couldn't compete with the doctor's cheerfulness, but he forced a smile.

A nurse appeared and detached him from his electrodes. Bill dressed himself, his mind still running with a hundred thoughts. Bess met him in his room as he was filling out his discharge papers.

"Dad, you better not be thinking of driving yourself home," she scolded.

He gritted his teeth. He loved his daughter dearly, but her worry was exhausting. He could say she got it from her mother, but Grace had learned to tone it down over the years. Or perhaps he'd learned that he better just listen to her. She'd go on strike if he didn't listen to her. He hated doing laundry. Grace had had a way of folding his socks just the right way. When she went on strike, he would cave, all over socks.

Pitiful, considering he now had to do his own laundry anyway.

"Honey, I'm just fine. And my truck is here, so how else am I going to get it home?"

"Well, if you'd called to tell me they were discharging you, I could have had Evie pick it up. That's what I'll do. I'll bring Evie to pick it up later. You're coming home with me."

"Bessie..."

"Dad, I'm not arguing. Grab your jacket."

"I don't have a jacket," he grumbled, and threw the pen down after signing the last of his discharge forms. He caught Bess's apologetic look at the nurse behind the desk. He took off for the elevators.

On the ride down, Bess paged through the thick pamphlet of discharge instructions they'd given him. "At least this doesn't seem too bad. We just have to get your blood pressure down. Have you been taking your cholesterol meds?"

"Yes," he grumbled, waiting for the elevators to hit the ground floor. Why were hospitals so big?

"We should go pick up your prescriptions."

The elevator dinged as it deposited them on the ground floor. Bill followed her out the hospital and into the sunshine. The fresh air filled his lungs—far better medicine than whatever the hospital provided.

Obediently he followed Bess to her car, looking longingly at his truck as they passed it. It was such a waste of fuel, having her drive him home. He was *fine*. When he slid into the passenger seat, she buckled his seatbelt like he was a kid.

"Really, Bess?"

"I'm sorry," she said quietly, and gripped the steering wheel with both hands. Bill looked over to see why she was waiting to start the car. Tears were welling from her eyes.

"Oh, honey." He softened and put a hand on her shoulder, then awkwardly pulled her towards him across the center console of the car into a hug. "I'm going to be fine, Bessie."

The tears were rolling freely now. "I know, but what if you weren't? What if you were out on the tractor, or throwing hay, and you didn't stop—I know how you keep pushing, pushing—and what if it was too late, and you were gone? Dad, I don't know what to do without you. With Mom gone, and Hector gone...you're all I've got left, Dad." She sobbed into his shoulder, wetting his shirt.

He rubbed her back. "Oh, honey. First of all, I'm not going anywhere. Not anytime soon. And give me a little credit, because I did listen to my old ticker and came to the hospital instead of finishing my chores. I had all night to question that. And Bess, when it *is* my time to go, you are not alone. You've got Evie, your sister Maggie, and Jeth, and Marissa, and all your friends. They've got your back as much as I do and don't forget it."

"But, Dad—"

"It's not the same. I know, honey. Believe me, I know. I love you something fierce, and you better remember that." His voice rasped at the last words, and he had to pull away to wipe tears from his own eyes. He took a shaky breath. "But apparently it's not time for me to go yet, so you're going to have to take me out to lunch and get me something better than hospital food."

Bess laughed, and the mood in the car lightened. She wiped her tears and started the car. Bill leaned back and watched his truck as they drove past it. Maybe he'd underestimated the family's recovery since Grace's passing. They were so solid, so brave. He sometimes forgot they missed her, too. And Bess, still raw from her husband's passing years back. He understood now how those wounds never fully healed. Maybe it was the wrong time to make a decision on the farm. For generations, it had been a consistent point for the family to return to, no matter the loss, no matter the arguments. Maybe they all needed the place just a little bit longer.

Chapter 25

Sienna burst into Jeth's office and came right to him, around his desk. Jeth fell back in his chair, hands in the air, as she pushed him away from his laptop.

"What the—"

"You need to see this," she said solemnly.

Was something wrong with Pap? He'd told Sienna about the hospital when he'd come in that morning, and she was concerned. But nothing having to do with Pap would be on his laptop. He leaned forward, watching as she pulled up the local news website, where the headline article sprawled across the screen was *Local Farmer sells out to Warehouse Developer*.

Sienna stepped back. Jeth pulled the laptop closer, reading fast.

"What!" His eyes scanned. The reporter had the story all right. RimForce, Inc. was proposing to build a million square foot warehouse—which was more than his company was planning for—in southern Schuylkill County. Farmer Bill Fogel's property was the land up for consideration. Local residents interviewed were shocked that such a project would happen in their backyards.

"I feel like Bill sold out on us," one neighbor said, according to the article. The Concerned Neighbor, who wished to remain anonymous, complained that a warehouse would detract from other local property values. It sounded like several of the people interviewed were planning on taking their protests to the township. "This shouldn't be allowed!"

Jeth finished reading and leaned back, then spun in his chair to look at Sienna, a finger over his lips. He pointed that finger at the computer screen. "It actually is allowed. The land has been zoned industrial for twenty years."

"I know. But now that the residents know, they're going to fight it."

"How on earth did that reporter find out about this?"

Sienna shrugged. "I doubt it would be someone in the company. This project is our bread and butter. When Clint finds out about it, he might put you in his crosshairs if things go south."

"Me? I've got more to *gain* than anyone if this project goes forward!"

"I know that. But he knows it's your family farm. The others don't—yet."

Jeth exhaled through pursed lips. "Who else would know? My grandfather's in the hospital. He certainly didn't call and sabotage this deal."

"Did you tell anyone else other than your family? Did Evie?"

"Crap—I wonder if Evie told Tyler."

"My first impression of the man wasn't good, but would he do something like this?" Sienna furrowed her brow.

"I dunno. The guy's changed since high school, if my cousin's broken heart is any indication. I think she told him. How could she not, if she convinced him to come over on a treasure hunt?" Jeth thought a minute, then pulled his personal cell out of his pocket and dialed Tyler's number. Tyler answered on the second ring. "Hey, Tyler, we didn't get to talk much the other day. What are you doing tonight?"

Sienna watched him as he played the game back and forth with Tyler a bit, settling on wings and beer at the local bar. He hung up.

He looked up at Sienna. "You want to go? It's actually a pretty good place. A bar, but not a dingy bar. And the food is awesome."

Sienna shrugged. "Sure. But what exactly are you playing at?"

"I'm going to get the truth out of Tyler. If he's sabotaging my family, it's time to find out why."

Evie pushed her glasses to the top of her head and wiped her brow. She leaned back against a teetering pile of books, which threatened to collapse behind her. She looked around her. One could easily be buried in this store. Good thing they didn't get earthquakes in Pennsylvania. All around her were piles of used books in the process of being sorted to be shelved. She stretched her legs out, snaking them between the piles. Her backrest hadn't collapsed yet, so she closed her eyes, breathing in the sweet smell of old paper and ink.

She'd been at this for hours, trying to catch up after her numerous days of missed work, which had started in the library and most recently been lost while shopping for a new car to replace her squashed one. Her assistant Marsha entered the back room, hesitating before delicately stepping between the piles to deliver Evie a fresh cup of coffee. The woman was worth her weight in gold.

"You're a lifesaver. Thank you," Evie said and took a grateful sip of the fragrant coffee.

"You have anything ready to shelve?" Marsha asked.

"Mmm. That pile there is good to go. And that one" —she pointed to a pile of torn paperbacks— "is ready for the crafter's corner." Why people thought it was okay to let books be eaten by mice was beyond her comprehension, but she didn't have the heart to throw any books out. Crafters would remove torn covers and making some amazing art with the remaining chunks of paper. They didn't mind if pages were missing, and the books were being put to good use.

Marsha scooped up the first pile. "Your friend Jayelle is here, too. Want me to send her back?"

Evie straightened and looked at the mess around her. "Um, I better come up front." With a sigh, she extricated herself from the maze of paper. Jayelle was in the main store scanning shelves, a few thick volumes already in her arms.

"Hey, lady!" Jayelle smiled when she saw Evie and gave her a hug. "How are you? Brilliant idea with the back-to-school section by the way. Pity I've already read half of those."

"You still expanding your library at home?" Evie laughed. It was good to have friends that shared her passion.

"Yes!" Jayelle waved the new titles in her arms towards Evie. "But anyway, that's just a side effect of being here. Beth and I are going out tonight, and we want you to come along. It's been too long."

"I've been kind of playing catch up—"

"We know, which is why we're abducting you. After work. No overtime today. It's wing night. I'll pick you up, say...seven?"

Evie rubbed her forehead, looking for more excuses.

"Pleeeease?" Jayelle pleaded.

It had been awhile since she'd gone out with the girls. She groaned. "Okay, fine. Seven though. And I'm not staying out until midnight. We aren't twenty-one anymore."

"Oh my word, like being twenty-five is so *old*." Still, Jayelle bounced on the balls of her feet, the dark curly ringlets of her hair springing with her. "I will see you soon." She pointed a finger to Evie and made her way to the checkout counter.

Evie was still standing among the shelves, a bit stunned, when Jayelle paused by the front door. "Oh, and dress cute. It's been a year, girl. Who knows who'll come out tonight." With a wink, she left, the bell above the door tinkling in her wake.

Chapter 26

"So I told him, if he wants it, then he needs to earn it," Jayelle was telling their trio of friends. They erupted into laughter at the bar, turning a few heads. It was still early, the bar far from crowded, but the change in scenery and conversation was soothing to Evie. Her friends were good to her. She'd told them about Pap, and they'd made it their mission to cheer her up. Mom had called to say Pap was back at the farm, doing well, so at least that was a relief.

She wasn't quite sure how she'd been finagled into picking up Pap's truck, but Jayelle had offered to help out there as well. Now that they were all two—maybe more—drinks in, they might be re-evaluating that.

"Evie told Tyler the same, I heard." Beth nudged Jayelle and giggled. "He tried to make another run for it, and she told him to get lost."

Evie rolled her eyes, her attention drawn back into the conversation. "It was far more complicated than that, but yes, we were done before, and we're even more done now."

"So now we have to find you a new man," Jayelle raised and lowered her eyebrows with a mischievous grin. She spun on her stool, eyes searching the room around them, with its dim lighting and dark wood paneling. TVs showed a variety of different sports, and a jukebox played in the far corner of the room. As far as patrons went, the crowd was sparse. "Hmm, let's see…"

"Jayelle, it's eight o'clock. Anyone here now is probably either with his wife or as much of a workaholic as Evie and needs to get to bed early." Beth joined her in scanning the crowd anyway.

Jayelle discreetly pointed to a classic lumberjack type of guy seated down the bar from them, broad shouldered and bristle-faced. Alone. "Bingo." Her lips formed the little O with amusement. Evie had to admit, the guy did remind her a little of Tyler's burly physique. Did she have a type? She debated for a moment, uncertain.

Beth snatched Jayelle's hand and pulled her around. "Oh, no-no-no-no." She leaned towards the other two girls. "That's Benny O'Hara. I work with him in the warehouse, and he's a total player." Evie and Jayelle giggled. Beth wasn't as amused. She whipped a ring out of her pocket, one of the rubber wedding band types. "You see this? I don't wear this sucker just for fun."

"You're faking that you're married just to avoid Benny O'Hara?" Evie asked.

"Shhh! He'll hear you," Beth hissed. It appeared to be too late. Benny looked their way, noticed Beth, and waved. She hastily slipped the rubber ring onto her finger, then gave a short wave back. She looked away, her back spinning to him as if he wasn't there. Her cheeks took on a pink tinge. Evie furrowed her brow, studying the man. He looked vaguely familiar.

"So are you faking because he's a jerk or are you faking because you actually like him?" Jayelle asked. She continued to watch Benny with a raised eyebrow.

Evie giggled into her drink—rum and coke, not beer—as she watched Beth's expression. "I think it's both!"

Beth rolled her eyes. "Look, the warehouse is good pay and all, but I'm a manager. I can't indulge in flirtations with any of my guys, or they aren't going to get their work done. Not to mention, I'd be fired. And Benny sometimes just doesn't shut up."

"So if you weren't his manager..." Jayelle grinned.

"I *am* his manager. Come on, you guys have no idea the pace they set for us in there. Like, who can unload a pallet of fifty-pound boxes in twenty minutes? And yet that's what some time engineer says we have to do it in. You know how much I busted my butt to get this promo? I'm not risking it for anything. Except maybe to quit my job entirely and go indie like Evie."

"What? And always wonder where your next dollar is coming from?" Evie scoffed.

"Hey, girl, you do darn well for yourself. *And* you're your own boss. Don't take that for granted," Beth said.

"I know. And I love my shop. Still, there's days where a nice consistent paycheck, with benefits, would be freakin' awesome."

Jayelle raised a finger. "Um, yes. I will attest. It's freakin' awesome."

Beth shook her head. "With bank hours like yours maybe. I work seven to seven most days, and in the winter that means I never see the sun."

"If you hate it, why don't you quit?" Jayelle asked with a genuine, curious tone.

"It pays well. For now. And someone's got to do it. I mean, who makes it their goal in life to work in a warehouse loading and unloading boxes?" She leaned on the bar with her chin in her hands. "We all like our little packages in the mail though. And it requires storage space to fill all those millions of orders all at once. Sheesh, my next dream promotion, if I was into data, would be as a logistician. Those guys are making a killing right now. They're amazing."

"So we need to find you a logistician." Evie smiled and threw her arm over Beth's shoulders.

Beth shuddered. "Ugh, no. Nerdy guys with glasses aren't my type."

"Right, she's way more into the lumberjack type," Jayelle said casually, making Evie snort into her drink. Jayelle broke out into laughter and looked over to Benny O'Hara, who was watching them with amusement.

Beth groaned, her ring finger still in plain sight of all. "I'm married, remember?"

Evie and Jayelle doubled over with laughter, the mirth rolling off them in waves until their ribs hurt. Only when Evie regained control of herself and wiped a tear from her eye did she realize how much she'd needed that laugh, not for the joy of the conversation, but for the release of tension it provided.

"I missed you ladies," she told them.

The door opened and closed behind them. Evie was still smiling as she took another sip of her drink, at least until she looked into the mirror on the back wall of the bar and saw Tyler's reflection there. His own expression seemed to fall into a cloud of gloom, and he wordlessly crossed the bar to the far side, past Benny O'Hara, and pulled up a seat.

"Is that—" Beth asked, her thumb discretely pointing.

"Yup." Evie downed her drink.

"Dang," Jayelle whispered. "What the heck are the odds?"

The door opened again behind them, and this time all three women looked up to see who it was. Jeth and

Sienna. They spun around to face them, various looks of interest on each of their faces.

Sienna instantly noticed Evie and smiled. Jeth's expression looked more complicated, which Evie thought was strange.

"Hey, lady!" Sienna came over and gave Evie a hug like an old friend. "I love that I keep getting to see you."

"How's it going, Sienna?"

"Oh, that's an interesting question." She glanced at Jeth. His eye was already down the bar, on Tyler. Sienna leaned forward and whispered in Evie's ear, "We'll tell you everything later." She straightened and waved at Evie's friends. "Enjoy your wings, ladies!" She followed Jeth down the bar. To Evie's surprise, they sat right next to Tyler, who looked like he'd been expecting them.

"Now what is going on here?" Jayelle said softly.

"I really don't know," Evie replied. "Judging by Jeth's expression, something is about to go down. But maybe not, since he brought his girlfriend along?"

"She looks like a Barbie Doll," Beth whispered. Evie couldn't miss the tone of scorn in her friend's voice. At one point, a long time ago, Beth had had a crush on Jeth. Apparently it hadn't completely died yet.

"Yup. She's actually super nice, too. And brilliant." Evie watched the introductions down the bar. "Do you all feel like we're spying?"

"I think Benny O'Hara is soaking up the attention. We have to look right past him to stare at the others. I give him three minutes, and he's going to come introduce himself." Beth ordered herself a new drink.

"That could make this even more interesting," Jayelle observed.

They watched as whatever conversation was going on with Jeth and Tyler intensified. Sienna seemed to placate them for a while. They ordered drinks. Benny O'Hara left his seat and approached the three girls.

"Hey." He leaned on the bar and smiled—at Evie.

Suddenly she realized why he looked familiar. Her expression brightened. "Oh my gosh, you're the trucker from the accident. How are you?"

"Good." Benny nodded. "You get your car squared away?"

"Yes, the insurance company came through, thank goodness."

Beth injected, her eyebrows raised, "Um—yous know each other?"

"Yes, well—no." Evie blushed slightly. "Evie." She held out her hand and Benny shook it.

"Benny. And I see you are friends with my boss." He shot a sideways smile at Beth. "Hi, Beth."

"Benny," Beth acknowledged, color rising in her cheeks.

Evie explained, "He was in the pileup that day as well. One of few drivers that didn't hit anyone. And we helped save a dad with two kids. Sheesh, that was a wild day."

"So you're a hero, huh?" Beth leaned her elbow on the bar, eyeing Benny with a bit more scrutiny.

Jayelle giggled into her drink and turned away.

Benny didn't have time to respond, as the attention of the entire room was drawn to the far end of the bar. Jeth and Tyler were both on their feet, fists clenched, and Sienna was extending a hand towards both of them to keep them apart. It was hard to tell who had shouted, but the volume between the two men was no low decibel.

"It was none of your business, Tyler! You should have kept out of it!" Jeth had fists balled at his sides.

Tyler leaned towards Jeth, only inches away from his face. "The residents have a right to know!"

Jeth hissed something back quietly, and Tyler didn't seem to be taking it well.

Evie pushed off her stool, pushing past Benny. He tailed her over to the men, but she was so distracted she didn't think to tell him to back off. She joined Sienna in pushing the two men apart. Sienna shot her a worried look.

"What the heck is going on?" Evie asked.

"Evie, this is under control," Jeth growled, his eyes never leaving Tyler's, who remained only inches away.

Evie glared. "I kinda doubt that, since you've got the whole bar watching. How about we all go outside?"

"I like her idea," Tyler rumbled and cracked his knuckles.

Jeth visibly bit his tongue, then stepped away. He motioned to the door. "Go. We're done here."

Tyler pursed his lips, then reached for his beer, downed what was left in one long swallow, and turned to Evie. He leaned into her ear and whispered not-so-quietly, "You think your family is above reproach. You're wrong. And tell your new boyfriend to bug off."

Evie could only furrow her brow in confusion. Tyler turned away, shoulder chucking Benny, who stood next to Jeth with his arms crossed, watching all of them. Benny staggered a step, his jaw dropping as Tyler shot him an icy glare.

Tyler turned and walked backwards to the door. He pointed at Benny. "Good luck. The tramp will play anybody."

Evie gasped, as did Sienna. Jeth crossed the room in two strides and landed a punch on Tyler's jaw. Tyler staggered back, only to be caught by Benny, who clamped a big hand on his shoulder and dragged him out of the bar. He threw him out the door and slammed it shut. Jeth shrugged off the now-angry bartender and apologized. The bartender retreated back behind the bar. Sienna and

Evie sank onto stools and were soon joined by Evie's two bewildered friends.

"What the heck was that about?" Beth asked with wide eyes.

"I don't even know." Evie looked to Jeth, biting her lip. Tyler's words had cut, though she didn't know where they'd come from.

Benny stood at Beth's side. "Hey, I didn't mean anything. Just wanted to say hello." He gave a short wave and returned to his meal at the bar.

"What the heck is going on?" Evie looked from Jeth to Sienna back to Jeth.

The four of them finally let their eyes settle on Jeth.

"No one talks about my cousin that way," he finally said, meeting Evie's eye.

Evie swallowed. So he did still have a brotherly streak to him, after all. "Thanks."

"What was the rest of it about?" Jayelle asked, her hand on Evie's arm.

Sienna met her eye cautiously. "I think you'll see on the news tomorrow." She inhaled. "Tyler told the press about the farm, Evie."

"No!"

Jeth and Sienna nodded.

Jayelle and Beth looked bewildered. Beth laid a hand on Evie's shoulder.

"Look, Evie, you guys have some talking to do. We're going to finish our wings. Please come back over then. Okay?"

Evie nodded and watched Jayelle and Beth stop to speak with Benny. She turned back to Jeth. "Tyler told them what exactly?"

"It's going to be a big stink. He told the press. They interviewed Pap's neighbors. They're throwing him under the bus, claiming all he cares about is the money, and nothing for the value of their properties."

Evie pursed her lips. She'd warned her mom the neighbors would feel that way. "Why would Tyler do that?"

"He's a jerk," Sienna cut in. She leaned against the bar. "I don't judge you. He's hot. But please tell me you now see through that man's manipulative nature?"

Evie swallowed. "So that's what the fight was about? Him telling the press?"

Jeth nodded. "He's trying to ruin the family. Or at least ruin the sale for us. He doesn't have the power to ruin the family. I don't know what he gets out of it. Maybe just preservation of the farm, because it's pretty. I was trying to find out; that's why we met him here. But things escalated more than I anticipated."

"A sick revenge, maybe." Evie sighed.

"Is the other guy even your boyfriend?" Sienna asked.

Evie snorted. "Heck no. He's just an acquaintance." She looked to Jeth. "You did more to defend my honor than he did. Though he did throw Tyler out—that was pretty impressive."

"Hmm." Jeth drank his beer. "Look, Evie, I'm going to try to salvage my date. And I owe the bartender a huge tip already, so I'm going to drink my fill. You and your friends good?"

Evie stood. "Sure. Hey, thank you both."

They nodded, and Evie returned to Beth and Jayelle, who still spoke with Benny in serious tones. She felt more than a little stunned. They greeted her with open arms.

Beth ordered her a shot. "I must say, your cousin has a bit of a big brother syndrome going on."

Evie forced a smile. "He's a good guy." She turned to Benny. "And speaking of good guys, thank you. That's the second time you helped me out."

Benny tipped his glass to her. "No worries."

Beth reached to touch Evie's arm. "I told him how you contacted your ex just to seek a geologist's opinion. He knows a guy at a local university."

"He's a good friend of mine," Benny offered. "Here." He fished a business card out of his wallet.

Evie studied the card and then took a picture of it with her cell phone. "Thank you. Now you're helping me out yet again."

"I'm hoping it puts me in my boss's good graces." Benny shot a sideways look at Beth and then straightened from his chair. "Well, you ladies enjoy the rest of your night."

As Benny worked his way to the door, Evie and Jayelle turned to Beth.

"You have to stop pretending your single around him. He's smitten. So are you. And he's a good guy," Jayelle insisted.

"I'll tell him if you want," Evie offered.

"Shh...no!" Beth looked aghast.

"Tell me you don't like him." Evie watched as Beth turned red, her lips sealed.

Benny was almost to the door.

"Hey, Benny," Jayelle called out.

Evie pointed to Beth. "She's not married."

Benny smiled, a hand on the door. "Yeah, I know. And I'm transferring to the trucking department so she won't have to worry about work relationships. Now she just has to decide if she likes me." Benny's eyes twinkled as Beth blushed a deeper shade of red, but her smile still crept into her features. Benny winked and stepped out of the bar.

"You two are terrible," Beth groaned.

Jayelle and Evie threw their arms around her. "We love you, too."

The girls bantered on for another hour, and still Evie could not fully relax. Her mind spun about what the night's events would mean for the sale of the farm. The barb Tyler had dug in tugged—*tramp*. She quit drinking, the buzz making her more queasy than happy.

Evie sat, a hand on her chin, staring into the lights across the room as her friends chatted on. They would be there a while. Jeth and Sienna stood to leave. Jeth left a wad of bills on the counter that was visible across the bar. He exchanged a few words with the bartender, then handed Sienna her jacket.

Evie acknowledged that her plan to relocate Pap's truck from the hospital needed to be reconfigured. As Jeth and Sienna approached the door at Evie's back, she apologized to her friends and excused herself for the night.

"Oh, you party pooper." Jayelle pouted. "Didn't you need a ride somewhere?"

Evie hesitated, then resigned herself. "Yeah, but I'll just do it tomorrow. You ladies have a good time. This was fun, but I'm done for the night."

They hugged her goodbye.

"Hey, don't be a stranger, okay?" Jayelle said and hugged her.

"And don't let Tyler get under your skin. You deserve way better." Beth smiled.

"Thanks, girls. Hey, get home safe!"

"Yup! Bye, Evie!"

Evie stepped into the cool fall evening, the noise of the bar fading as the door closed behind her. She breathed deeply for a moment, then swung her keys on the fob as she headed to her car. Jeth and Sienna were across the parking lot, making their goodbyes. She smiled to herself and turned away. She unlocked her car and tossed her purse inside.

The rumble of an engine fired up across the parking lot, and Evie looked up as Sienna drove by in her Jeep. They waved to each other. She looked back to Jeth, who watched her from outside his truck. Evie shut her car door and walked over to him, hands in her pockets, shoulders hunched against the cold.

"Hey, you don't want to go with me to drop off Pap's truck, do you? My mom wouldn't let him drive, and Jayelle was going to take me but I don't want to interrupt the girls' fun." Evie kicked at a stone in the parking lot.

"I really just want to go home, Evie."

"Okay, fine." She shrugged and headed back to her car. She'd expected as much. He had a lot going on. Still, he'd thrown a punch at his former best friend in her defense. Didn't that merit conversation? She turned back. "You know…"

Jeth was halfway into his truck and didn't hide his frustration. "Now what Evie?" he snapped.

She froze in place. "Sorry." She turned on her heel and beelined it back to her car. She'd wanted to thank him, to tell him how much it meant to her, but his tone stung. She'd overstepped her bounds, crossed some invisible line, probably by ruining his date for the millionth time. Her nerves raw, she gritted her teeth against tears. At her car, she turned and yelled across the parking lot. "Sorry to be the annoying cousin, Jeth!"

She crawled into the front seat of her car and slammed the door. She leaned her forehead on the steering wheel. *Tramp. Annoying cousin.* A truck rumbled in front of her, then stopped, its engine idling. She could feel the vibration of the diesel engine through her seat. Jeth opened her car door, and she hastily wiped tears.

"Evie, what are you talking about?"

"I'm annoying. I can't get along with anyone. I ruin everything."

"Knock it off, Evie. Get out. I'll take you to get Pap's truck."

"No, I'll do it myself. I'll take my bike in the morning." She fumbled with her keys.

Jeth reached across her and plucked the keys from her shaking hand. "No, we're going to get Pap's truck. And you're going to tell me what all this emotion is about. You aren't acting like yourself, Evie."

Evie felt the flood break. She avoided Jeth's gaze, instead staring at the steering wheel, the bar, anything. "You have your mom, your dad, and your sister. And now Sienna. Who do I have? My dad is gone. I never had siblings. My mom is trying her hardest to hold herself together. Pap is—I was so worried. Tyler—I thought he was it, Jeth. I really did. And now I'm like gum on the bottom of his shoe. All I've got is you, you big oaf." The tears were streaming now. "Don't you treat me like I'm some annoying cousin. You are my *brother*, whether you like it or not." She turned away from him, fury and frustration bursting forth in a flow of tears and snot. She wasn't helping her case.

"Get out, Evie." His words were so gentle she couldn't help but obey, the car door clicking shut behind her as she stood and wiped at her eyes. Then Jeth's arms were around her, his firm grip enclosing her in a way only Pap and her father could mimic. It wasn't a hug that asked anything. It was the *Hey, I've got you, you're safe* kind of hug. She leaned into him and let the tears flow, then eventually pushed away and wiped her nose on her sleeve. She'd left a wet spot on Jeth's shirt, which she brushed at before giving up.

"I'm sorry," she whispered.

"Evie, you are family. You are never going to be alone. You need to know that. Okay? That will never happen. Not in this family."

Evie nodded.

"You can always call me. Or my parents. Or even Marissa. And then there's a circle wider than that which we can't discount. Our family is blessed to have each other. And that's how it's going to stay. Even when Pap is gone."

Evie's chin trembled again.

"But, hey, he ain't gone yet. So take it a day at a time, okay? Everything may seem like it's happening all at once, but this is all small stuff."

Evie again nodded, then took a deep breath, letting it out slowly.

Jeth put his arm around her and steered her to his truck. "You know, for what it's worth, I love ya like a sister, too. I think that's what happens when you put cockroaches in someone's sleeping bag."

"That was you? I always thought that was Charlie! I didn't talk to him for months. You jerk!" She punched him lightly in the shoulder, pulling away. He opened the door for her to climb into his truck, and she locked her car with the press of a button. They'd pick it back up on their way back from Pap's.

"It's hard enough to picture losing the farm, but Pap too?" Evie said quietly as Jeth slid into the driver seat.

"I know. That place is like home. Is it Pap or the land that makes it that way?"

"I think it's the memories. Generations and generations of memories."

"Memories we keep in our head though. No one knows what physical object ties into the memory until someone talks about it. That's what makes home. That's what binds family." Jeth pulled onto the road, heading towards the hospital for the second time in two days.

"The objects remind us though," Evie said quietly, thinking of the arrowhead back at her house. It wasn't just the story of the land and the Indians before them that was important. It was the story of her and Jeth up on the hill, prospecting for gold that didn't exist. It was the story of their mothers taking friends up there as teenagers. Memories of the past creating the present that influenced that future. "Jeth, can I borrow that rock you found on the hill?"

"No."

"Don't be a jerk. I'm trying really hard not to be annoying."

"I want to show you the real deal."

"Not happening."

"Suit yourself. No rock."

Evie sighed and slouched back in the seat. The lights of town came in greater frequency now, an economically successful town's boom evident in every bright sign.

Once she transferred to Pap's truck, Evie continued to think the entire ride to Pap's house, which wasn't very long. There were a lot of changes happening in their little community, and a lot of them were for the better. They'd fixed the highway after decades of talking about it. They'd built the new hospital. The downtown area was turning into someplace you wanted to walk around, with two coffee shops, a library, hairdressers, heck—even a yoga studio was downtown now. Was all that development coincidence? Or was there a "they" behind it all with a greater plan, to prep the area to become the next hot area for development?

Evie's eyes narrowed as she passed the sign for the new gas station, a ten-acre lot for sale commercial, another new chain store, an overgrown field with a sign claiming lots for sale, and then the open expanse of cornfields, with Pap's driveway winding down to the single streetlight in the old barnyard. Evie made the right, and stopped in the middle of the driveway. The glittering lights of the highway raced in the not-so-far distance. A billboard glowed in its high-tech neon colors behind her, giving the view an eerie glow in the twilight. And then there was Pap's house, with its solitary light.

It stood like a beacon in a sea of concrete and asphalt. Could this really be the last farm?

Chapter 27

The news broke across the small community the next day with the same intensity as a murder. Not that murder ever happened in their little town. People entering Evie's shop talked about it non-stop, completely unaware that she was a granddaughter of the famed farmer.

"Can you imagine? A warehouse that big right in our backyard?"

Evie bit her tongue and kept busy. She *could* imagine a warehouse in Pap's backyard. He would no longer have a backyard. No tire swing. No barn. No neat rows of corn. These women would not lose anything. They lived across town, in the ritzy section, and wouldn't even be able to see the warehouse from there, even with its monstrous size.

By mid-afternoon her nerves were fried. She called her mom, worried that Pap was going to be hounded by the town. Someone had even claimed the township was holding an emergency meeting that night in the borough building.

"What do you mean they're having a meeting?" Bess exclaimed into the phone. "We have full right to sell! It's all in the zoning, perfectly legal. What limits they impose on building the thing are the responsibility of the developer, not us!"

"I know that, but the town seems to not care. They've been talking about it in the shop here all morning. Does Pap know that word is out?"

"No. Maybe. How could he?"

"Mom, we have to go tell him. If he finds out on his own, it might stress him out—"

"Okay, honey, I hear you. I'll go over now. When will you be done work?"

"At five. Well, a little later with closing."

"Okay, come over then. I'll make dinner. And I'll call Jeth. I think we need a family meeting to discuss how we ride this out."

"Agreed." After Evie hung up, she had deep empathy for Jeth. The poor guy had a girlfriend, and here he was getting called back to the farm every night. Though, that would happen even more if he bought it.

The day went by quickly, and Evie was able to confirm that the rumors were true, and an "emergency" township meeting had been called to discuss the development plans. When she pulled into the farm, Jeth was still not there.

Bess had the table set and a roast chicken with veggies finishing in the oven.

Pap was out on the porch, bundled in his winter coat with a hot cup of coffee. A bite of fall was in the air. He didn't look happy. Evie slipped into the chair next to him wordlessly.

They sat in silence awhile, listening to Bess hum and clink dishes together in the kitchen behind them.

"I feel like an old man," Pap finally grumbled.

"You've got a ways to go to be old, Pap."

"Everyone making plans around me, like I'm not here."

"Not true. You get to make all the decisions. I don't envy you that, either."

Pap sighed.

"Where's Jeth? He comin'?" Evie peered down the lane.

"No. He's going to the meeting."

"So Mom told you?"

"Yes. All these years living next to these people. None of them lifted a finger to help me maintain this place. Help came from much further way, from Chuck and you and Jethro and Charlie. But not from my neighbors. They were so wrapped in their own lives." He shook his head sadly. "I can't blame them, either. I can't. I did nothing to help them either. But now they sabotage me as if I'm bulldozing their front lawns. This—" he waved his hand across the fields, their little green rye plantings pushing forward. "This was my work. My yard. They may have enjoyed the view, but they did nothing to own it."

Evie inhaled. She could feel the pain of betrayal in his words. "I'm glad Jeth will be at the meeting."

"He said he had to be there anyway. For work." Pap sipped his coffee, then threw it into the yard with a deft movement of his wrist.

Evie's eyebrows shot up in alarm.

"Cold." Pap rose on stiff knees, then offered Evie a hand to her feet like a gentleman. "Good thing dinner's ready. And your mama has turned into a darn good cook, if the smell of that chicken is any indication." A hint of his old smile returned.

Evie took his hand and pushed to her feet.

The meeting was chaos, and Jeth didn't want to be there in any capacity, much less as the bad guy. And no matter what role he played, he was the bad guy. The crowd was out for blood, the arguments heated, the rep from RimForce overwhelmed despite his years of professional experience. The only thing going for Jeth was that no one had yet figured out that he was the grandson of the farmer.

"That family just wants the money!" someone shouted. "Corporate greed!"

Corporate greed it may be, but what did that have to do with a farm family?

The township officials were trying to restore order. They tried to explain that nothing illegal was happening. That all the rules and regulations allowed this sort of building, and that they would all be followed to a T. That permits were being filed in full compliance. They tried to explain the benefits of big business coming to their area, but the residents could not be quieted.

Charlie appeared at Jeth's side. He looked up at him with wide eyes.

"Hey," Jeth said quietly and shook his friend's hand.

"Hey." Charlie shoved his hands back in his pockets. Jeth noted the faint smell of cows. Who was milking tonight and feeding calves?

The arguments went on.

"Cutthroat, aren't they?" Charlie whispered.

"Yup. I'm so glad Pap isn't here."

"Me too. Poor guy." Charlie chewed his lip. "Hey, if he wants an out that isn't a warehouse, I'd like to try to buy the place. I just can't pay developer price. But I could get the whole operation rolling again. That property has the space for it."

Jeth turned his full attention to him, ignoring the debate. "What about your family farm?"

"We're next on the development list. And my brothers want to sell. They saw we can buy a place with robotic milkers and three times the herd in Iowa, all for what they offered us for our land."

"Charlie..."

"I know. I'm furious. My kids are established here. I'm established here. My wife's family. Heck, my family."

"That farm is your life."

"You're telling me. All they see is the dollar signs though. And they're right. We could make more with fewer hours labor if we upgrade. To Iowa." He gritted his teeth. "Just wait until this crowd hears there's a second farm going down."

"Are there going to be any left?" Jeth mused.

"Golden question, right?" Charlie pursed his lips. "Hey, I've seen enough. Please, talk to your grandfather for me. I'll be looking into all my options. Okay?"

"I will, Charlie. I will."

Charlie slapped him on the back and stepped out. Jeth longed to follow him, but instead stood waiting along the wall, waiting for his boss, Clint, to introduce him as the engineer designing the evil warehouse itself.

Jeth drove up Pap's lane late that night and was glad to see the lights were still on and cars were still outside the house.

Aunt Bess met him at the door with a big hug. "Hey, honey. How did it go?" she whispered.

"Not good," he whispered back. "Pap asleep?"

"Yes, thankfully," Evie said quietly. "He's acting like he's fine, but he really is exhausted."

"It's good he didn't go tonight. The residents are like wolves circling, and there's no pleasing them."

"I can't blame them," Aunt Bess said. When Jeth and Evie both looked at her with surprise, she added, "Imagine building your dream house on five acres overlooking this picturesque farm, only to find out that you'll now be overlooking the roof of a warehouse."

Jeth sighed.

"You hungry?" Aunt Bess asked.

Jeth nodded, and she wrapped an arm around him and guided him to the kitchen.

As she heaped a plate full of leftovers and stuck it in the microwave, Jeth leaned his head in his hands. "I really wish I could just buy this place."

Silence echoed around the room.

"You really feel that way?" Aunt Bess crossed her arms, her eyes studying him.

"I feel like it's my calling."

"Those are strong words, Jethro."

"I know."

"Then act on it. Figure out how you're going to do it. Make a plan."

"What can I offer?" he looked up at his aunt, the tortured back and forth of every conversation he'd had about the place running through his mind. Offer more. Offer less. Quit his job. Use his job to pay off the mortgage.

"You offer what you can. You decide how you want to spend your time. Then you sit your grandfather down, and you let him decide."

"But what's fair to Evie? And my sister?"

"They will get a share of what you pay."

"Do I have to pay what the developer is? I just don't know if I can do it...I don't want the bank to own it. If something were to happen and I couldn't make the payment..."

Aunt Bess came up and put her hands on his shoulders. "Jethro, you offer what you can. Simple as that."

Jeth inhaled sharply. He looked to Evie. "You didn't find a gold mine yet, did you?"

She smiled. "You're the one that found the legendary lightning rock."

"Come up and see it."

"I told you Jeth, I'm not going up there. You can bring the rock down. I'm not going up there again."

"But Evie, you need to see the whole thing."

"Every time I go up that mountain, it tries to kill me."

"Not every time. We were up there without incident a few weeks ago."

"And then it tried to kill me again. No. I'm not going." Evie folded her arms across her chest. "Jeth, what are we even doing?"

"What do you mean?"

"What if there's gold up there?"

"Well, if there's gold, maybe Pap won't have to sell the farm at all."

"You realize that only works if we buy the mineral rights to the land, which can be expensive?"

"So then RimForce has to pay more, too."

"Pap could lose the entire sale. They could back out." Evie was pulling at the hair of her temples again.

"Kids, calm down," Aunt Bess shushed. "Evie, what do you want your grandfather to do? Would you be okay if Jeth owned the farm?"

"I keep flipping back and forth on what I want Pap to do. He deserves a good retirement. He's worked so hard his whole life. And yet...to picture all this gone?" She looked around, then turned back to Jeth. "You're designing this thing. Will the house and barn at least be saved?"

Jeth shook his head. The project wanted the warehouse central to the property, which meant the house and barn had to go. Sienna's job was to determine how to re-route the creek in an environmentally sound way.

"They'd tear down a two-hundred-year-old building?" Evie asked with sad eyes.

"They might sell the wood to be repurposed. Maybe not."

"They might not reuse it?"

"They could sell the whole barn. They did that with a barn down in Elizabethtown. They moved the whole thing. Now it's a wedding venue."

"Moved it?"

"Yes. But it was unique architecturally. It cost them a fortune to move that barn. Our barn is just a barn."

Evie sighed and sank into a chair across from Jeth. She looked to her mother, then back to him. "Jeth, do you like your job?"

"Not lately."

Evie pursed her lips. "You know, life is pretty short to work a job you hate."

"Do you like your job, honey?" Aunt Bess asked Evie.

Evie nodded. "I love it. Some days it's a job, as all jobs are, but I love the inventive, entrepreneurial aspect of what I do. I love seeing people's faces light up when they find a new favorite book. I love interacting with the community." She turned to Jeth. "I think I get that from Pap. You remember when he'd work the farmer's market booth? He knows everyone in this town." She narrowed her eyes. "What do you love?"

Jeth looked at his hand, flexed his fingers. He loved the challenge of his job. Loved the numbers game. And yet he also loved working outside. It was when he felt most alive.

Looking at his fingers, he realized how proud he was of his callouses. Most of all, he loved it when he was doing something different every day.

He looked up, his eyes meeting Evie's and his aunt's. He wasn't sure if he was describing engineering or farming. Maybe it was both, but he was only one man. He had to make a choice, and he had to do it soon.

Chapter 28

Sienna was on a conference call when Jeth walked into her office. It looked like an intense one, if her throwing her stress ball from hand to hand was any indicator. She held up a finger to silence him before he could even speak, and he leaned against the closed door. She took the call off speaker and held it to her ear, swiveling in her chair, turning her back to him.

"Look, I don't make the laws. But I do know we need a bog turtle study on that site before they destroy the wetland." Sienna listened to whoever was on the end of the phone, then suddenly held it away from her ear, wincing. When she replaced the receiver her tone was the cold tone of a riled boss. "We do the study like the law says. I'm not risking my reputation by taking shortcuts. If they can afford to build, they can afford to pay for a wetland survey." After another moment's pause, she hung up and spun to Jeth, scowling.

"That sounds fun," Jeth said with a sarcastic smile.

"Totally. What's up?"

"Lunch?"

"Not today, I'm afraid."

"Dinner?"

Sienna sighed and glanced at her phone, calculating. "Maybe. I'd like to. Depends what time of day you want to eat though. It's amazing the number of hours we have to devote to one turtle species. Not saying that they aren't important, mind you. But you'd think we'd look at every species in a development area, not just one turtle." She went back to throwing the stress ball hand to hand, watching Jeth. "It's for you grandfather's farm, you know. They're lining up the surveys as if he's already sold. They really want to move fast on this project. Any news if he signed yet?"

Jeth shook his head. "Kinda silly if we're doing all this work and he's not selling, isn't it?"

Sienna shrugged. "I guess it'll be done for the future. When someone sells it in another thirty years."

"Who'd sell it in thirty years?"

"Just theorizing. So that I don't feel like I'm wasting my time on bog turtles." She eyed Jeth carefully. "How is he doing?"

"Tired. His age is catching up."

Sienna hesitated, the ball poised for a throw that didn't come. "Did you make a decision yet?"

"No. He was in bed when I got out of that blasted meeting last night." Jeth crossed his arms. "Look, you sure you don't want lunch?"

"I can't. I'll let you know how things are looking for dinner though."

"I want to go visit Pap tonight, check up on him. You okay for a late dinner?"

"Sounds perfect."

Jeth winked at her. "It's a date."

He ended up eating lunch at his desk. By the time he looked up at his clock again, it was time to quit for the day. Sienna was back on the phone with another call, so he waved to her and then headed out of the office. Hoping to maximize his time with Pap, he took 78 back to Schuylkill County. On impulse, he started counting the warehouses lining both sides of the highway like rows of fifty-acre sized houses. Most weren't even labeled with a company logo. They were probably leased out. Trucks pulled in like little ants. The volume of moving trucks was nothing compared to those parked in the lots surrounding the warehouses, sitting stagnant.

What kind of infrastructure needed this much interior space? There was a lot of talk about sustainability lately. He didn't see how this was sustainable. How many warehouses did a company need? The streetlamps surrounding the warehouses—currently off during the day —had to number in the hundreds. Pap had *one* for his entire farm. Jeth didn't care how many LED's were used—they still used electricity.

Jeth tried to picture what the land had looked like before the bulldozers came. The rolling fields on the next stretch of highway gave some indication. Picturesque farmhouses and silos sat in the valleys that had not yet been developed. This land south of the Blue Mountain was flatter than up in Schuylkill. He didn't see any evidence of them having had to blast away mountainsides to level the ground. Still, they certainly had moved a lot of dirt to create the flat land needed for the warehouses he'd passed. On his left yet another site sat under construction.

Were they building like this everywhere in the country?

Jeth used the voice feature on his cell phone to run a search, just to check a crazy thought.

"How many warehouses are for lease in Berks County?"

The robotic voice replied, "There are currently twenty-seven warehouses for lease in Berks County, ranging from three thousand to one million square feet."

His jaw dropped. He hadn't even crossed the Berks County line yet. Millions of square feet of farmland bulldozed. Billions into the local economy with jobs, including his own paycheck. Development was good, right? He'd gone to school for engineering because it was cool to see an idea come to life, to emerge from numbers and become a computer image, and then to be steel and concrete and functional space that more amazing things could happen within. Still, twenty-seven vacant buildings, occupying hundreds of thousands of square feet? That did not include the footage under roof that was already in use.

Jeth carefully passed semis through a construction zone, yet another infrastructure project to accommodate the trucking boom. His nerves still prickled just thinking about Evie's accident every time he drove this highway. Thankfully he made it to the Hamburg exit without issue.

It was a relief when he pulled into Pap's quiet driveway. Pap was sitting on the front porch with a glass of iced tea. He smiled when Jeth walked up.

"Hey, kiddo," Pap said cheerfully.

"Hey, Pap. Glad you're home. And taking a break for once."

"It is kinda nice. Your sister called, and she gave me strict orders to sit on the porch and drink some tea. She's a very smart lady, your sister. She's gonna make a fine doctor."

Jeth laughed and plopped into the chair beside him. The view was incredible on this sunny evening, with a few hours still to go before sunset.

"You need help with anything, Pap?"

"I think your buddy Charlie's coming over to get the apples sprayed. Not sure how we're going to mix the spray, now that it's been sitting in the sprayer separating out for two days."

"You got a long stick?"

"I suppose that would work. Just don't get any of that on you."

"I know, Pap." Jeth stood. "Maybe an old broom handle in the barn?"

"Should be one in there. I can look for you."

"No! You're resting." Jeth pushed down on Pap's shoulder, settling him back into his chair. For once he obliged. "Charlie coming tonight?"

"Yup. Probably when they're done milking. After dark."

"Late night for him. Should I just run the sprayer?"

"Naw, wind's still a little strong. It'll die off. That's why we spray at dark." Pap narrowed his eyes at Jeth. "First rule of farming is that you farm by conditions. Not a clock. Or a calendar for that matter."

"Got it." Jeth mock-saluted, and headed into the barn. It didn't take long to find the broken handle of a shovel, plenty long to fit into the big five-hundred-gallon spray tank and stir. He plucked a pair of rubber gloves out of the box in the workshop, just to be careful, and got to work. When he'd finished, he did a walk around, checking to see what else would have to be done. Pap was pretty good about putting equipment away. His organizational skills for the tools and other supplies were non-existent, but it had always been that way. Jeth realized the best way to figure out what would be needed next would be to have Pap make out a calendar of what he did when.

He returned to the porch, went inside to locate Pap's calendar on the wall and a pen, and went back out.

"Alright, Pap. I know we can't farm by a calendar, but you're gonna have to give me at least an estimate of when to do stuff. Pretend I'm going to be a farmer."

Pap didn't show emotion, and Jeth wondered why his joke hadn't held. "Where do you want to start?"

"January. That's pretty quiet, right?"

Pap laughed at that at least. "Nope. That's pruning season. And snow plowing season. The township doesn't own this road. That's on us. And I usually order my seeds around then. And we're always selling hay through the winter. Good hay prices around then."

Jeth jotted all this on the calendar. "Anything else?"

"Firewood, if you're going to run the woodstove. Haven't bothered with that in years though."

"Oil's expensive."

"So is time."

Jeth nodded his head in agreement. *Fair enough.* "So February..."

They continued like that through the calendar to fall, when Jeth started running out of room to write down all the things that needed to be finished, harvested, planted, winterized, and put away in the few short months of September through November.

"How on earth do you do all this on your own?" he sat staring at the page.

"Well, I get help. When you were in high school it was great. And I still hire the high school and college kids when I can, like you saw the other day. Though it's been hard to get help this year. I don't make fifteen dollars an hour, much less enough to be able to pay that much to kids that work at half my speed. There's a reason they call it the harvest moon." Pap looked nostalgic. "When my dad was a kid, he and his brothers would all line up in the field to bring in the harvest. One to work the mule, another to stack, another to drive the cart, with a second horse if they had it. They'd work all night in the moonlight, trying to beat the impending weather. First thing my dad did once he saved up some money was buy a tractor."

"Massey."

"Massey's grandma." Pap smiled to himself, looking out over the fields. "I think she's still sitting in one of the fencerows somewhere."

Jeth tapped the pen on the calendar. He tried not to feel overwhelmed. They finished out the lists for November and December, and by the end Jeth was feeling as if he'd been given a mission. He looked back to September, the things that needed doing now. A lot of it was mechanical prep of the equipment for harvest season, which they were getting into the heart of.

"How far did you get on the equipment, Pap?"

"Massey needs a hydraulic pump. It's patched for now. Going to have to get one fabricated in a machine shop since they don't sell the part anymore."

"You want to show me what you need? You feeling up for it?"

"Sure." Pap downed the last of the tea and rose.

As they worked, Jeth's mind wandered. Charlie must think he was a fool for letting this place go. He was a fool for not having taken an interest sooner. They just hadn't thought that far ahead, as to what would happen when Pap retired.

His father had insisted he go to college, and he didn't have to work hard to convince Jeth. He may not have gotten in on any football scholarship like he wanted, but the drive to go to a big university, be part of something of that scale, had been assumed his whole life. Plus, he loved the challenge of engineering. Math had always been his strong suit, so why not get paid well for it? There had never been a conversation about going to school for agriculture. And Marissa, his little sister, had been playing doctor since she was five years old. It was natural for her to commit her life to it. In just a few more years he'd have to call her Dr. Marissa Fogel. Ugh, he wasn't looking forward to that.

Evie was the wild card. Her reputation in town was well earned. She was a creative, shrewd entrepreneur that had created a great asset for the local community. With her business degree, she could have taken a job anywhere, in any field, and yet she'd ended up working for herself, running a bookstore in an age when some said bookstores were dying out. She'd prove them wrong. Now why didn't *she* go into agriculture? Jeth supposed there could be some creative ways to do it, but the physical work was overwhelming, as he'd been reminded when he'd thrown straw bales around the other night. Any farmer needed help. A female farmer, as sexist as it sounded, needed even more help.

Or one heck of a physical conditioning program.

They tinkered away in the barn until dark, when Jeth heard a truck pull up. He glanced at his phone. Seven o'clock. He'd have to hurry to make his date with Sienna at eight.

"I better get moving, Pap, before Sienna yells at me."

Pap reached behind him into the toolbox. "Better not mess it up with that one, Jeth. A girl like that doesn't come around every day."

Charlie's truck door slammed, and he came over to them. His eyes had bags under them, and he looked exhausted.

"Hey, Mr. Fogel. Jeth." He gave the sprayer a look over and seemed satisfied. "Just the usual apple trees, right, Mr. Fogel?"

"Yes, Charlie. Thanks for your help."

Jeth followed Charlie to the tractor, watching as Charlie gave the sprayer a look over. Satisfied, Charlie stretched and removed his ball hat to scratch the top of his head.

"Thanks for helping out," Jeth said.

"Anytime, Jeth." Charlie still seemed down, his usual smile lacking.

"So—how are things going?"

Charlie looked down, his jaw tight. When he spoke there was a bit of a waver to his voice, the emotion evident. "My brothers want to sell out. They've all but picked out the curtains on a place in Iowa. I offered to buy the farm, but Laura and I just can't afford it. Not in comparison to what the developer is offering. Even if I bought out my brothers at that price, the monthly payments would cripple us."

"You going to move to Iowa with them?"

"No, I'm not going to uproot my family. The kids have friends here. Not to mention family." Charlie sighed and crossed his arms. "None of this is going to happen fast. Maybe within the next year. We'll have to sell equipment, sell the herd." He snorted. "Sixty years of genetics, just disbanded to the wind."

Jeth felt Charlie's pain like it was his own. "I'm sorry, Charlie. That sucks."

Charlie was quiet, his chin quivering, then he took a deep breath and met Jeth's eye. "The good part is, I *will* get a third of the sale. Mom and Dad made sure of it. I'm going to look locally, see if I can find a smaller property. I'll have to do some kind of boutique farming, maybe organic." He shook his head. "I've been making fun of organic farmers for years. Never thought I'd be considering trying it. They can make about the same salary with a smaller scale though. And without my brothers, I'm going to need smaller scale."

"What's wrong with organic?"

Charlie shrugged. "Nothin'. But it costs a lot more to do it. Impossible to work economies of scale to your benefit. As soon as you get a sick animal that needs antibiotics, even if it's for something minor like an infected cut, you have to get it out of your herd for good. You can't spray your crops, so the yield is smaller. The feed then costs more. There's a compounded effect. That's why organic charges more. I'm not sure the income is a whole lot different compared to traditional farming. That's why a lot of people do it as a side job or hobby."

"Is that an option?"

"I already talked to our local ag supplier. They offered me a job in the nutrition department when I'm ready.

I have an animal husbandry degree and a lifetime of experience."

Jeth watched his indifferent expression. "You like the actual farming though, huh?"

Charlie nodded. "Sixty years of genetics, Jeth. Every two years or so a new chance to improve the herd. In my lifetime, I've already bred at least ten generations of cows. I've watched the improvement in the herd. I've been doing this since I was a kid in 4-H. And now I have to watch all those cows sold? I—I can't even talk to my brothers right now."

"What's your parents' take on this?"

"Dad says it's our decision. He wants to retire, so he can't force us one way or another. I think mom's heartbroken though. She doesn't want us to move to Iowa. And she's not really ready to leave her home either. But, if the price is right..."

Jeth raised his eyebrows and let out a whistle.

Charlie slapped his shoulder. "Well, I've gotta get goin'. Tomorrow starts soon enough. Take care, bud."

"You, too, Charlie. Thanks again."

They waved a last time as Charlie drove the tractor towards the orchard. If a family like that, actively farming, could fold, what chance did the Fogels have?

Jeth glanced at his watch again. Time was running out. He better not let Sienna down. He shouted a quick goodbye to Pap and headed out.

He made it in time. All through dinner, he stared at the beautiful woman across from him, discussed work a little, discussed her friends a bit more, including her best friend's upcoming wedding. And then they discussed the farm. A lot.

Chapter 29

E vie sighed. She'd searched the fields and stream for rocks like those Jeth had brought down from the mountain to little avail. A pile of rocks sat before her on Pap's porch next to a bucket of muddy water, which she'd used to clean them. "Well, this geologist dude is going to laugh at me if I show him these. Anyone can see that these are just...rocks."

Pap shrugged. "You can go up and look on the mountain."

Evie shot him a dirty look, which only made him laugh.

"The weather report says today is going to be a beautiful day. Sunny and seventy. Not going to be many more days left like this, so I suggest you get out and enjoy it. Go for a walk. Make your peace with that mountain."

Before it's too late, her mind added for Pap. "Lightning only strikes twice, right?"

"I think the saying goes 'lightning never strikes twice.'" Pap smiled.

"Well there goes that reassurance," Evie mumbled. She leaned back on her hands, legs stretched out in front of her on the porch. *Be a big kid, Evie. Go face the scary mountain while the sky is blue.*

With a sweep of her arm, she slid the rocks to the side of the porch next to the banister, then jumped to her feet. She went into the house and filled a water bottle from the kitchen sink. On a last impulse, she added an apple and a bag of chips to an old backpack. When she turned, Pap was leaning on the side of the doorframe, his arms crossed and a sparkle in his eyes.

"You want a compass and a knife, too?"

Evie stuck her tongue out at him and he laughed.

"You'll be fine, kid. I'll send a search party if you aren't home by dark."

Evie pulled the backpack onto her shoulders. "Come rescue me if a sudden hurricane whips up again."

"Will do." He stepped aside to let her out of the house, and she jumped off the front porch and headed across the field.

He was certainly right about one thing—it was a perfect day for a hike. There was just a little bit of a breeze, the sun was warm, and the leaves were changing colors, the yellows and reds still mingling with a heavy amount of green. When she got into the woods, the ground was a carpet of gold, the smell sweet. She followed the trail she'd taken Tyler up by, wondering what on earth she'd been thinking inviting him back into her life.

He was a jerk. He always had been. And she was a fool for not having dumped him long before.

Thankfully, it had been silence from him since the day Jeth had punched him. That was probably a good thing, because if she ever heard from him again, she would be giving him a very large piece of her mind.

Evie was so lost in her thoughts that it took the sharp smell of pine to make her look up. There before her was the poor tree that hadn't dodged the lightning bolt. It was splintered, blackened—a stark reminder of the power of nature to destroy even the strongest things. Before the tree was a broken boulder. Evie knelt by it and picked up a shard of rock. It was laced with flecks of yellow. Evie took a few of the best samples, as much as she could carry in the pack, and then straightened, looking up at the tree. The blue sky shown above it, the wind gentle. No threat of storm. She reached out a hand, placing it on the splintered wood. For a moment, it was as if she and tree understood one another, their fears at having to stand tall, even when afraid. They understood the importance of roots and recognized their fragility. And when she again looked to the rocky slope the tree grew from, she recognized that however difficult it was to plant those roots, if you were determined enough, you could always find a way.

An idea came to mind, and she shouldered the pack and continued up the mountain. She kept her attention sharp, looking for the minor landmarks she'd noted when Jeth had guided her. There was the flat rock she'd rested on. There was that funny tree with a twisted trunk. Soon she crested the ridge, and on the other side was the ruins of the little cabin.

Evie slipped the pack of rocks from her back and took a drink of water. She circled the ruin, studying it in more detail. The wood was long rotted away. Only the foundation had been stone, and un-mortared stone at that. It was old, certainly. She went into what once was the doorway, and poked through the leaves with a stick. Nothing to see. Outside the ruin, she dug a little deeper into the hole Jeth had found the coin. Just dirt. She surveyed the land around her, and then strode over to the dry streambed. She sat on a log and took another drink of water.

What had it been like to live here? A fresh spring right outside your front door must have been nice. The woods were quiet, peaceful. She could only imagine what it sounded like without the roar of engines, however distant they may be now. The threat of conflict with the traveling bands of Native Americans must have been nerve-wracking. It took a great deal of courage to raise a family this deep in the wilderness, at least the wilderness of those days. What had her ancestor's reaction to the natives been? Had they tried to trade with them and be friends, or did they stake their lands and instigate the aggression shown to them? Had they even been here at all, or was she sitting, musing, on some long-forgotten stranger's land?

She sighed and leaned back, looking through the colored leaves at the blue sky. She closed her eyes, breathed in the earthy smell of the forest. Without fear of lightning, she had no qualms with the woods or the mountain. A rustle of leaves alerted her, and she opened her eyes to watch a squirrel hop through the leaves. He had no fear of her, and she watched him a long while, until her arms grew stiff. As soon as she moved, he hurried up a tree and disappeared behind it.

Evie brushed the hemlock needles off her hands. She hesitated, looking at the ground she had disturbed. Then she reached out, her heart pounding, and picked up the rock. Sure enough, her eyes were right, and it was another arrowhead. This one was chipped, as if broken in the making. She spun to her knees and pushed leaves aside, searching the bottom of the old creek bed. She found another chipped arrowhead, and then another.

She looked around, incredulous. Even if this wasn't a village, this *had* been an encampment. She looked up. Hickory and walnut trees, which her research had told her the Indians sometimes planted near their favorite hunting camps, circled the area.

If you are determined enough, your roots will find a way.

Evie took a final look around the ruin, then headed down the mountain, her thoughts swirling. Hiking tours, archeology lectures, camping, corn mazes, pumpkin launchers, maybe even horseback trail rides—she had to find someone to run that—oh, or maybe hay rides. The ideas flicked on like Christmas lights, each strand making the last look even brighter. The core was that the family add options if they got into agritourism. If Jeth wanted to farm, and Evie could run her creative business layout, they might just be able to make it work.

Even if Jeth was her most annoying cousin, he was still her favorite. And they just might be able to work together.

Bill looked up in surprise from the kitchen table when he heard the thud on the porch. Evie was visible through the screen door, the bag of rocks deposited next to the others. She pushed through the screen door, letting it slam behind her. Her hair was a wild halo around her, wind tossed, and she was breathless with a sheen of sweat on her forehead. Had she run down the mountain? There was no storm, the blue sky still visible through the door behind her.

Before he could open his mouth to ask, she blurted out, "Agritourism."

Bill blinked at her.

"Agricultural tourism," she said again. Then she stepped up to the table and placed a handful of rocks in front of him.

Bill reached for his glasses, then looked closer at what she was showing him, expecting to see gold-flaked rock shards. Instead, he saw arrowheads, a whole handful of them. He looked up at Evie with wonder.

"Where?"

"By the ruin of the cabin. Right next to that old creek bed. And there's walnut and hickory trees."

"All of these? In one place?"

"Yes! Don't you see? There's a historical landmark here. It could be preserved. And with the main farm, there are so many fun activities we could host to educate the public on farming. Or just let them enjoy the fresh air. And if Jeth seriously wants to farm—"

"I haven't heard if that's Jeth's plan—"

"But if it is, Pap...we could then keep this place." She looked at him with big, earnest eyes.

He looked back to the pile of dirty arrowheads she'd set before him.

"Oh, honey."

"Pap, we have options. So many options. A warehouse is only one."

Chapter 30

The pulse of the music vibrated in Jeth's core. It felt good to move on a dance floor on a Saturday night with a beautiful woman, particularly when that woman was smiling with her arms twined around him. Sienna looked stunning tonight, not that she usually didn't. She was in her element here though, an elegant but well-fitted dress hugging the right places, her makeup just a little more dramatic than she wore for work. She was a good dancer too, which made him feel a little like he had two left feet. He hadn't stepped on her toes yet though.

He took her hand and spun her in a circle, and she laughed, the note sounding over the music that rocked the big converted bank barn that was serving as her friend's wedding venue. The bride danced her way over towards them, a little bit of a cha-cha in her step. She pointed at Sienna.

"Don't you disappear, lady. The bouquet toss is next!" She continued her cha-cha around the dance-floor, rounding up her friends.

"I never liked those things," Sienna said to Jeth, her hands caressing the back of his neck as their hips swayed. "The girls are either climbing over each other or no one wants to try to catch it."

Sure enough, the music soon changed and the DJ called out for all the single ladies to come to the dance floor. Sienna snagged a glass of champagne from a passing waiter and downed half of it before handing it to Jeth. "I'll just hide in the back."

Jeth smiled, sipping the rest of the champagne from the edge of the dance floor. Beyoncé blared over the speakers.

"You better listen to the words to this song, dude," the husband of one of Sienna's girlfriends said to him. "Girls like that don't wait forever. Trust me, I married one."

Jeth laughed. The bride got ready to throw, dramatizing it. True to her word, Sienna stood to the back, but Jeth saw a certain determination in her posture. She was a little lighter on her toes, like a runner ready to take off for the next base. The bride threw, the ladies stretched out their arms, and the bouquet landed neatly, a perfect shot, right in Sienna's hands. She stared at it as the crowd cheered, and then with a blush looked up at Jeth.

The man next to him clapped him on the back. "That says it all man. Now if I were you, I'd be certain I caught that garter."

"They're doing a garter toss?"

"Of course. And I hear they're not abridging it. The man that catches it has to get it up the bouquet-girl's leg—all the way up—so the bride and groom have good luck. You don't want another man running his teeth up your future wife's leg do you?" He laughed and walked away.

"Wait—teeth?"

The DJ called for single men to come to the floor.

Sienna stepped up next to him, blushing furiously. She held the bouquet lightly in her hands. "Sorry?" she said bashfully with a little shrug.

Jeth kissed her on the forehead, then put his game face on. The groom removed the garter from the bride's leg, then spun it around on one finger. "I didn't think they still did this sort of thing," Jeth mumbled to the man next to him.

"You're in PA Dutch country. We've got all the old traditions. I hear there's even a pig trough for the groom's older brother to dance in."

Jeth shook his head, but when it came time to line up, he made sure to place himself in the front of the group. Some of the men seemed a bit eager to meet the pretty bouquet-winner. The groom stretched out the garter, aimed it over his head, and let it fire. It was going high, and Jeth's old football instincts kicked in, and he jumped. He caught it and tucked it like a game-winning pass.

Beaming, he turned back to Sienna, who blinked at him with sparkling eyes from behind the bouquet. Was she smiling? Oh, yes, she was smiling. It was almost too obvious, as her friends noticed and cheered and pulled her back onto the dance floor and into a chair. She crossed her long legs at the ankle and tucked them to the side, the toe of her high-heeled stiletto just grazing the floor. Her short dress didn't leave much protection, which judging by the clamp of her thighs, she was well aware of. She sat with her back as straight as a princess, the bouquet lightly placed on her lap. Her eyes met Jeth's with a spark of challenge.

"Now the idea here is to get the garter up as high as possible. For every inch, the bride and groom get another year of good luck," the DJ was saying to Jeth through the mic. "Now, if you want to double that number, you've got to get it there with your teeth." He clapped Jeth on the back. "Good luck to you!"

The crowd hollered all around them. Sienna kept her eyes locked on Jeth's, her lips in an embarrassed smile.

He knelt before her, the garter in his hands.

"You know, the bride is one of my best friends," she said quietly.

Jeth looked up at her, then placed the garter in his teeth. One hand gently untangled one of Sienna's legs from the other, and she pressed her skirt down with the bouquet. He pulled off her shoe—no sense getting stabbed in the eye with a heel—and then kissed her foot. With only a moderate amount of difficulty, he looped the lacy garter around her foot, and in one smooth motion, had it halfway up her thigh. That was when it got trickier. The crowd was shouting encouragement, Sienna was pulling at her skirt as his nose brushed up the skin of her leg. He got it to the hemline, then let go with his teeth. Before she could protest, he slipped it under her skirt with both hands, and made sure it was *all* the way up.

He quickly stood and offered Sienna a hand, pulling her to her feet as the DJ mused on just how high he'd gone, but they weren't paying attention to the crowd. He could feel Sienna's pulse through the contact of their fingers, see the way her breath had quickened. His own heart thudded in his chest.

He kissed her fingertips.

Soon others were joining them on the dancefloor, the party resumed.

"Come for a walk with me," Jeth asked quietly.

Sienna nodded.

He led her outside the warmly-lit barn and onto the groomed stone path that led to a small propane firepit a short distance away. The firepit blazed cheerfully, but the wedding goers were all down in the barn. They had the place to themselves. They walked hand in hand, then sat on the bench by the fire. Sienna shivered in the cool fall air, so Jeth draped a conveniently located blanket over her bare shoulders, then hugged her into his side. She rested her head on his shoulder.

"I'm sorry I haven't been around much. And I'm sorry I've been short with you," he whispered.

"You have a lot going on. I understand."

"You deserve more though. You have a life going on as well. Tonight has made me realize that. You did great helping your friend with this wedding."

"Thanks, Jeth. But really, what I have going right now is nothing compared to what you've been thrown the past few weeks. I'm grateful you've given me the time you have."

She kissed him on the cheek, then curled back into his shoulder.

"Sienna..."

"Hmm."

"That day in Pap's kitchen, when we said we could try to buy the farm together and then just sell if it didn't work out...I think Aunt Bess was right. We shouldn't be doing that if we aren't sure we're getting married. I'd feel like I'm using you."

Sienna was quiet. "I was just trying to be helpful."

"I know, and I appreciate it." He was ruining the moment, putting his foot in his mouth. How could he voice that he didn't want her father to take the credit for buying his family farm? How could he voice that he wanted her to be married to him before they moved in together? He wasn't quite ready to propose, but her friend's husband's words rang true. He knew he wanted her. He knew she was a kind, smart, and beautiful person. He just didn't feel he was certain enough about his own life to take that step. Didn't she have the right to know if she was marrying a well-off engineer who could give her a beautiful, champagne filled wedding and a big home, versus a farmer who very well may end up broke trying to preserve a lifestyle?

To Sienna's credit, she wasn't taking it personally. She wasn't pulling away. They watched the fire a few minutes, and then Jeth whispered, "I love you."

She looked up at him, her eyes sparkling. "About time you actually said it." She leaned up and kissed his lips, then cupped her hands on either side of his face and swung her leg over his lap. The blanket fell to the ground. Jeth sat stunned, at her mercy, until she suddenly pushed back to her feet and offered him a hand.

He needed a few moments before he could return to public.

"Why do you like me?" He tipped his head to the side.

"Because you're a good man, Jethro Fogel-Mueller. You're honest, and you may be a bit thick headed sometimes, but I know you try your best to take care of the people around you." She smiled. "You're kinda cute, too."

Jeth took her hand, gave her fingers a light squeeze, then stood. They both straightened their clothes, then arm in arm returned to the party barn and danced the rest of the night away.

Chapter 31

The man with the Tesla was there again on Monday. Bill threw down his rag and stalked over to him. The harvester was going to have to wait.

"I didn't sign yet," he growled.

"Are you planning to, Mr. Fogel?" Todd Banks asked, a folder tucked under his arm. He absently straightened his tie, his expensive suit spotlessly clean.

"I just got out of the hospital," Bill growled.

Todd blinked. "I'm sorry to hear that, sir. Are you feeling better?"

"I'm supposed to be keeping my blood pressure down."

"Sir, I tried to call again. RimForce needs an answer before they start pursuing other possibilities."

"I didn't answer because I was in the hospital." Bill faced off with the man, his hands on his hips. But the salesman didn't budge.

Instead he handed Bill a card. "How about you give me a call tomorrow? Once you've had time to relax a bit. Then we'll talk."

"Whose farm are you going to try to buy next?"

"Excuse me?"

"You keep saying you're going to move on to the next guy. Who is it?"

"I can't disclose that, Mr. Fogel."

"There's only one farm left, and its Chuck Hess's. His three boys are all farmers. They ain't gonna sell."

"Mr. Fogel, if the price is right, anyone can be convinced to sell. Now the offer we made you is high because of your location here. It's convenient—"

"They ain't gonna sell. They've got two generations of kids coming up the ranks, and they still are farming like they've got mouths to feed. Which for the record, this country does have. Mouths to feed. Once you've built warehouses on all the farms, where are you going to import all the food from? Overseas? The fertile ground is here! You might be able to relocate our textile industry and our manufacturing and our steel plants, but you can't relocate our farms."

"Mr. Fogel, I—"

"If we close our farms, what will we eat?" Bill pointed at the bumper sticker on the back of his pickup. "Why isn't the US government paying better attention to that, hmm? Instead they're giving tax breaks to mega-corps to develop areas. That might be fine in a desert, but we're on fertile ground here. This soil is some of the best in the world, and it's been tilled for centuries. Even the Indians in this area farmed!"

"Mr. Fogel!" Todd shouted, his face red. He held up a hand to Bill. Bill quieted, scuffing the dirt with an angry kick. Todd's voice softened. "I don't disagree with you. In fact, I understand better than most. That's why I keep asking to talk with you. My parents were farmers. They sold. They bought an even bigger farm for less than what they made from the sale."

Bill stilled. Part of him wanted to shout *traitors*, but another part of him was genuinely curious about this development. He hadn't thought about buying another farm.

"Mr. Fogel, is there someplace we can sit down?" Todd looked concerned.

Bill had to remind himself he'd just gotten out of the hospital. He sighed and gestured towards the house. Bill followed him.

"May I ask why you were in the hospital?"

"Heart attack," Bill said shortly. Hopefully the man would take the hint.

He didn't comment.

Bill led him inside, the screen door slapping behind them. He set a cold glass of tap water in front of them both, then took his seat. Todd sat across from him.

He didn't touch his water.

"Believe me, I understand what a difficult decision this is. My family farmed their original homestead for nearly two hundred years down in Chester County. We loved the history of the place, but all around us things were being developed. More houses, more shops. All nice stuff. But the country suddenly wasn't the country. They were given a very good offer for the place, and they took it. They ended up buying another place in the western part of the state. More acreage. And it's truly in the country—back to driving thirty minutes for a grocery store again. But they love it. It's how it used to be for them." Todd leaned across the table. "Mr. Fogel, the same thing is happening in your area. The cities are expanding. The economy is growing, which is good for the town. There are conveniences with these projects you may not even realize. Better power reliability, better internet, better roads, better hospitals. RimForce helps fund these things both through our substantial tax contribution and our own infrastructure projects. But it leaves you on an island in the middle of it all. That island has become very valuable, Mr. Fogel. And RimForce is willing to pay far more than any private developer, and certainly more than any farmer, could pay. It would be enough for you to continue farming somewhere where you don't have to be on an island. If you choose."

Bill took a long drink from his glass, weighing this. "I think I can finally picture how the Indians felt as the settlers encroached west. They were offered a price, and suddenly it seemed reasonable to just sign over the land, and move a little bit more west." Bill leaned towards Todd across the table. "They were pushing into the land of others, and even then it wasn't as good. Dry, rocky. They knew what they lost, but for those eastern tribes, it was too late. I won't debate the fairness of the treaties signed, but is any contract really fair when there's no going back?" Bill leaned back again, looking down his nose at the man across from him.

Todd was quiet.

"My family has been stewards of this land since the Indians left it. This is a weighty decision for me. *I need more time.*"

The suited man nodded. "I will tell RimForce you've had some health concerns. But there is only so much time I can buy you before they move on, and the offer will be lost. And you still will be left on an island."

Bill met Todd's eye. "Is it such a bad thing to be stuck on a green, fertile island, when the sea is made up of concrete?"

Chapter 32

C lint pushed open Jeth's door so hard that it smashed against the frame holding the window in place. The glass shivered, but held.

"In my office. Now." Clint stalked away.

Jeth scrambled to save his work and pull up his lock screens, then hurried out. Sienna was following Clint into his office, her spine straight. She shot a worried look at Jeth. Once Jeth had crossed the threshold, Clint slammed his door with enough force to make the glass shudder again. Sienna flinched. He stood with his hands on his hips, facing them.

"You two are dating."

Jeth and Sienna looked at each other with wide eyes. It wasn't exactly against company policy, but generally speaking, interoffice relations were frowned upon. They'd kept things professional in the office. They *were* professionals, after all. Jeth turned back to Clint and met his glare.

"Yes."

Clint clapped his hands together. "I knew it!" He started pacing, hands back on his hips.

"Is that a problem, sir?" Sienna asked.

"Yes. I won't have you both being distracted. Messing around on *my* time!" He stopped pacing and pointed back and forth between the two of them. "One of you has to quit!"

"Excuse me?" Sienna's jaw dropped.

"You can't do that," Jeth protested.

"Then break up." Clint leaned against his desk and shrugged.

Jeth looked over at Sienna. He could practically see the smoke coming out of her ears.

She turned on Clint, pointing her finger in his unshaven, pale face. "This is harassment. Nothing in company policy says *anything* about interoffice relations, thus this is none of your business. If you try to fire either one of us, we have grounds for a wrongful termination lawsuit."

"That's why you have to quit." Clint smiled at her, and Jeth bristled at the intensity of that look.

Sienna exploded. "You will not have a say in my personal life. My relationship with Jeth has stayed outside of the office. We are your two most productive employees. We meet every deadline. You have no grounds for this."

"I own the company. If I don't like it, I make it go away. And I don't like this." He again pointed back and forth between the two of them.

"Suck it up!" Sienna shrieked.

"Break up!" Clint yelled back.

"Enough!" Jeth yelled above them both. He was surprised at the power of his own voice. It had taken on a tone he hadn't known he was capable of. "Enough." He looked at Sienna. This multi-talented, assertive, kind woman was it. He held no more doubts. "Sienna, I love you. I want to spend, way, way more time with you. Will you marry me?"

"Are you freakin' kidding me?" Clint's jaw dropped.

Sienna's eyebrows were in her hairline. "You mean it?"

"I was going to go ring shopping tonight, so I apologize if things are a bit out of order."

"Yes! Yes!" She flung herself into his arms and hugged him, then kissed him.

"Ugh, I'm going to have to fire both of you," Clint grumbled. He sank into his chair.

When Sienna pulled her lips from Jeth's, she smiled sweetly at Clint. "You can't afford to fire us, and you know it."

"Just one of you, please, quit. I can't watch this disgusting, sappy—"

"I quit." As soon as the words left Jeth's mouth, he knew he'd made the right choice. Sienna looked at him in awe. Clint looked like he'd swallowed glass. "I. Quit."

"You're joking?" Clint asked.

"That's what you wanted, isn't it?" Sienna asked sweetly. She whispered in Jeth's ear, "I hope you have a backup plan."

He kissed her on the cheek. Hand in hand, they left Clint slumped in his chair.

Jeth pulled her into his office. He closed his laptop and left it and the majority of the contents of his briefcase on his desk. He scanned through his work phone, then set that aside as well. Feeling considerably lighter, he smiled at Sienna.

"First, I'm going to go buy you a ring. I hope it's okay if it's a little diamond, because then I'm going to use every last penny I have to buy a farm. Are you okay with that?"

Sienna threw her arms around him. "I love that plan."

He pushed her back to smile into her eyes. "Keep your job, Sienna. Poor Clint's going to need you. Because I'm also going to strike off on my own, get my own engineering firm started. I know a lot of clients that are going to switch over in a heartbeat, because they're tired of being sidelined by the big corporations."

"Maybe I'll work for you," Sienna shrugged.

Jeth kissed her forehead. "Ultimate power couple, huh? Let's make sure we can afford a farm first."

❧ · · ◆ · · ☙

Bill dialed the number of the lawyer in Philadelphia, Sienna's father. The phone rang once and a sweet-sounding secretary answered.

"Law Offices."

"Hello there. May I speak to Attorney Vonn? It's Bill Fogel."

"One moment please." The line clicked to boring hold music, then with a rattle, someone was on the other end again.

"Bill! I've been meaning to call you. I looked over the contract, and everything looks squared away. No hidden meanings. I'm actually rather impressed with how well it's written. They've got a heck of a legal team on their side."

"So I could sign it?"

"You could. Honestly, it seems like a good deal. Congratulations."

Bill thought for a moment. "Mr. Vonn, did you know my family has owned this farm for over two hundred years?"

"That's amazing."

"Yeah, it is."

The silence stretched.

"Well, Mr. Fogel, I am going to have to get back to work. Do you have any further questions?"

"No, sir," Bill said hastily. "Thank you. Hopefully we can meet in person in the future. My grandson seems to be quite smitten with your daughter."

"Ah, well, if half of what she says about him is true, he's a good man. Maybe we will. Good day, Mr. Fogel."

"Thanks again." When Bill hung the phone back on the wall, he stared at it for a long time. There was nothing in the contract that he couldn't or shouldn't agree to. He took off his cap and ran a hand through his hair. It was thinner than it had been in his youth, yet another reminder of how he'd aged. He was running out of time. It was time to plan for the next generation.

He walked the few steps down the hall to his office and slid into his desk chair. He opened the file, and stared at the signature line. Maybe he should tell Bess and Maggie before he signed. Just to get their feedback one last time. He stood again and walked back to the hall phone.

He dialed Maggie first.

She picked up right before the fourth ring. "Hi, Dad," she said brightly. "How are you feeling?"

"A lot better, sweetheart. How's Florida?"

"It would be better if you came down to visit. Rick's out fishing now. You should see the sea bass he brought in last week. You'd love it, Dad."

"I might do that."

"Really? When?" He heard her excitement through the phone.

"Maybe next week?" He paused, then spit the words out. "I've decided to sell, honey."

"It's the right decision, Dad. I know it's a hard one. But you deserve this for your retirement. Who knows, maybe you'll like Florida, and you can move down here by us."

Bill smiled. "Maybe."

"Well, thanks for letting me know. Can I start looking for a plane ticket for you? I'm so excited for you to come see our new place. I can't believe you haven't been down yet. We'll go fishing! And you should see the river where the manatees swim. It's beautiful."

"Sounds great, honey." If Maggie was this excited, it had to be the right decision. There were so many vacations he needed to catch up on. "I have to call your sister, yet. I'll talk to you more soon."

"Love ya, Dad."

"Love you, too, sweetheart." They ended the call. He dialed Bess, feeling much more confident in his decision. Bess picked up right away.

"Hi, Dad. Everything okay?"

"Why does everyone ask that when I call?"

"Probably because you were in the hospital only a few days ago." Bess laughed. "You sound okay."

"I'm fine, sweetie. I'm going to go to Florida next week."

"Oh? I thought you had so much work to do?"

"I do. But…" He took a deep breath. "I'm going to sign, Bess. It all will be theirs, so the planting isn't going to matter much."

There was silence on the end of the phone.

Bill frowned. "You there, sweetie?"

Bess cleared her throat. "Uh, yup. It's your decision, Dad."

"What's wrong, honey?"

"Well…Jeth was talking about making you an offer. Did he change his mind?"

"He's been talking about it for a while, but I'm not sure if he's serious." Bill scratched the top of his head. "A salesman from RimForce stopped today and said that they're going to move on to the next landowner if I don't make a decision soon. And I spoke with the lawyer, and he said the contract is good to go. Don't you think this is the right decision? It's a lot of money, Bess. I'm not going to spend it all in my lifetime."

"Dad, I support you fully with whatever decision you make. I do think you should call Jeth one more time though."

Bill sighed. "Alright. But I have your approval?"

"Either way." She paused. "I love you."

"Thanks honey. I love you, too. Why don't you come over later?"

"How about I make dinner again?"

"That would be amazing, Bess."

"Okay, see you then, Dad." They ended the call, and Bill was back to feeling conflicted again. He hesitated with his hand over the phone. If Jeth wanted the farm, what would he do? Could he accept less in order to keep it in the family? Was that fair to the other two grandchildren? He took a deep breath and dialed his grandson's phone number.

It went straight to voicemail.

Jeth looked at his phone as he left the jeweler, a beautiful diamond ring in a little box in his pocket. The phone was dead. He groaned inwardly and looked around the mall. There was an electronics store several shops away, so he navigated his way there around the moderate crowd of people. He bought a charger, and when he got back to his truck, he plugged in the phone. Maybe quitting his job was a bad idea. Ring, farm—now a phone with a good battery life was on his list. He was hemorrhaging money. It was going to work out, right?

Once the battery was up to three percent, he turned it on. Almost immediately, the dings of text messages came through. Ten text messages from Evie. A missed call from Pap. And a text with a single heart from Sienna. He smiled to himself at that one, then turned back to Evie's, bracing himself for whatever drama she would throw at him next.

CALL ME. CALL ME. WHERE ARE YOU? went the list of messages from Evie. Jeth inhaled and turned on his Bluetooth earpiece, then dialed. As the phone rang, he pulled out of the parking spot.

"What the heck, Jeth! I've been trying to reach you for an hour."

"It's been an eventful day. What's the emergency?"

"Pap is selling. My mom just told me he decided today. He was trying to call you, too."

"What?" Jeth slammed on the brakes as someone backed their car up in front of him. "You sure?"

"Call him!"

Jeth pulled into the newly vacant parking spot, his hands shaking as a rush of emotions flashed through him. "I guess he has the right to make that decision. I can't offer what RimForce was offering."

"Well, if you don't want the land, then he made the right decision. I tried to talk to him about agritourism, but he's retiring and I can't handle that big of a venture on my own. I'm already busy with the shop."

"Evie, I just quit my job. And proposed to Sienna. I'm making changes."

"Whoa." He could hear her surprise through the phone.

"I thought you liked Sienna?"

"She's freakin' awesome. Congrats, Cuz."

"Thank you."

"Now what about the farm? And why the heck did you quit your job?"

"Long story. I'll call Pap. Where are you, anyway?"

"Almost to the university. I'm meeting that geologist Benny told me about. He actually sounded interested in what we found. We'll see. At least I'll close that chapter of drama. And never have to think about rocks again."

"Good luck, Evie."

"Thanks. Same to you."

She hung up, and Jeth immediately dialed Pap. The phone rang. And rang. And rang. Pap didn't believe in answering machines. Jeth tried his cell phone number, which went straight to a voicemail box that was full. He gritted his teeth in frustration and hung up. He was headed to the farm next. He could only hope that Pap had decided to take another tour around on the tractor, and had not yet signed.

Chapter 33

Carrying a backpack of rocks across a college campus was a little awkward. The pack was small, but she had a feeling that she walked like an old lady. Who carried around a backpack of rocks?

She looked up at the science building and pushed through the revolving front door. Then she began the laborious process of carrying the backpack of rocks up two flights of stairs.

The geologist that met her in the lab could not have been more different than Tyler. He was about Evie's age and held the classic look of a nerd, with thick glasses, messy short hair, and a button-down shirt with more than a few wrinkles. When he looked up at Evie and smiled, the excitement lit his face like a kid on Christmas. It was undoubtably genuine. That bare honesty brought a handsomeness to his features that caught Evie off guard. She swallowed.

"Hey, Miss Evie. I'm so glad you could come in." He held out his hand and she took it, his grip firm and gentle at the same time. "I'm Doctor Gregory Hudson. Just call me Greg." He let go of her hand and pushed his glasses up his nose.

Evie pulled the heavy backpack straps higher on her shoulders, her heart racing. She was a little hot, likely just from lugging rocks up the stairs. Right?

Greg was still talking rapidly, his excitement evident. "This is probably the coolest thing I've heard about in my career. I wrote a paper as an undergrad about the possibility of gold in this area of the country, and always hoped someone would come forth with more evidence. My doctoral profs told me it was impossible. I'm so excited to see your find."

She blushed as she slid off the backpack. "We weren't sure how many samples you needed." She rubbed her shoulders after she handed it over.

Greg hefted the backpack and laughed. "Didn't quite need this much. Where on earth did you park?" He took out the first fragment and squinted at it, then carried it over to his microscope without waiting for her to answer. "Wow." He studied some more, then reached for a tool with a tip as fine as wire. He prodded the rock, then set the tool back down, still staring into the microscope. "I can't believe it. Check this out." He waved her over to the microscope, where Evie could see the rock placed under the light.

He gestured for her to take a look, and Evie obeyed. The rock on the other side of the lens was beautiful. The glassy black was streaked with slivers of yellow that seemed to have melted like currents of a stream through the rock, each line finer than fishing line.

"What am I looking at?"

"A geological miracle. That's gold. In a sedimentary fulgurite."

"Gold? In a what?" She looked closer.

"I've never seen anything like it. Not in Pennsylvania. I mean, we do see gold in all kinds of rock. Sedimentary, igneous, metamorphic...even granite and quartz. This rock is laced with quartz and gold, the particles rearranged by intense pressure we usually only see at asteroid impact sites. You said this boulder was on top of a hillside?"

"Yes."

"Likely a glacial deposit," Greg said to himself. He peered through the microscope again. "It takes considerable force to rearrange the particles into deposits visible to the naked eye. The force alone—"

"It was a lightning strike." Evie shrugged as he spun towards her.

"That would do it. A bolt powerful enough can actually rearrange the atomic structure of the rock. The shock lamellae—" He stopped himself, smiling as he took in Evie's wide eyes. He pulled off his glasses and looked her up and down, fidgeting with them in his hand.

"Bottom line is, I've never seen anything like this before. Though considering it exists, it does a lot to explain where the placer gold in the rivers around here comes from. You see, gold *can* occur anywhere. But certain conditions mean it is more common some places than others. This is unique, Miss Evie. These trace amounts of gold are so far into the rock, I doubt they would ever be found unless extreme energy, like a lightning strike, split the rock to show them."

She looked back to the backpack and laid a hand on one of the rocks inside. Tyler was wrong. He could have seen it, had he given her the time of day and looked. "So the Indian legend is true?" Evie whispered to herself.

"What legend?"

"Of the Gobbleberg. 'When lightning strikes, the rocks are cleft in twain, and gold can be found within.'"

"Never heard of it, but that sounds right. How old is that legend? Wish I'd have heard of it."

"It's from the days when the Native Americans were still in these hills. Apparently we found the Gobbleberg." She didn't know what to make of his enthusiasm. Evie removed the rock from the microscope and hefted it in her hand. "So what do I do now? Can we mine this?"

"My suspicion is that these boulders are too few and far between to make it worth a commercial enterprise. But that doesn't mean that your farm wouldn't be a prospector's paradise. You'll have to file the paperwork to protect the site, but then you can control who, even if it's just family, can look for these rocks. The refining process is going to be rough. You almost have to replicate the heat of the lightning to smelt out the gold ore. It's going to take some equipment."

"So we aren't going to be rich?" Evie's face fell. She's had a small degree of hope, but what did it matter? The farm would be sold.

"Probably not. But—this is really cool." The scientist squatted down just enough that she was forced to meet his eyes. He pushed his glasses back up his nose and straightened. I'd love to come see where you found it. Maybe I can help you determine what you're sitting on. If it's enough, maybe you even want to charge amateur prospectors a fee to poke around. You know, like they do at some of the local caverns."

"You think we could have a tourist attraction?" She raised her eyebrows with skepticism.

"Yes!" He nodded enthusiastically.

Evie shook her head. It was all pointless. The farm was for sale. And if the gold was too hard to get, it didn't stop or change the sale.

The geologist stilled. "Why the long face? Isn't this what you wanted to hear?"

Evie looked up at him, emotions fluttering through her mind. "It is. It just doesn't change the future like I hoped.

My grandfather is selling, and we need way more than un-refinable gold to save the farm."

Greg shoved his hands in his pockets, looking crestfallen. "The site is for sale?"

Evie nodded, avoiding his gaze.

The geologist sank onto a stool. "I have to tell you, Evie. This is the find of a century in the geological community. If word gets out about this, if you contact the local papers, tell them about your old legend, the place is going to double in value. You're going to be getting calls from geologists all over the country who want to research the phenomenon."

"My grandfather's already getting well over asking price. How much more could it be worth?"

"Who's offering over asking price?"

"A warehouse developer."

Greg shook his head. "You can't let them bulldoze a find like this. It's too rare."

"What can I do about it? Deny my grandfather his retirement?"

The geologist ran his hand through his hair, making it stand on end. He almost looked charming that way. "Can I see it? Before the sale?"

Evie shrugged. "He was going to sign today. If it's not too late, you can look."

The geologist perked up again, his eyes twinkling. "I'll get my coat."

Evie set the specimen from the microscope on Greg's desk. "Hey, keep this one. For your help."

"Thanks." He picked up the rock once more, smiling at it, then tucked it into a drawer of his desk. He pulled on his coat and withdrew keys from the pocket. He held out a hand to Evie. "You want me to carry your bag down to your car for you?"

Usually Evie would protest, insisting she was an independent woman. Today though, she could tolerate a little chivalry. "Sure." He took the bag with ease and pulled it over one shoulder, the strap stretching precariously. She led the way back through the building.

"Don't you have classes this afternoon?"

"No, just papers to grade. They can wait."

Evie pressed her hands into her pockets, and her hand brushed the arrowhead. "Hey, maybe I should show you this, too." She withdrew it and handed it to Greg as they walked. He turned it over in his hand for a moment, then stopped.

"Come with me." He turned and went the other way down the corridor, knocking on a colleague's door. The man, apparently an archaeologist by the look of the office, turned the piece over in his hand with a studious eye.

He turned to Evie. "You found this locally?"

"Schuylkill County. My grandfather's farm."

"It's an exceptional specimen. Probably Lenape. Has the site been further excavated?"

"No."

The man sat on the edge of his desk. "I've been pushing for protection of some of the Delaware tribe's hunting grounds north of Virginville. The Lenni Lenape had a village to the south, but seasonally they would migrate north, over the mountain. They would have temporary camps as they went, usually in roughly the same places. They are hard to find, since there usually were not too many things left behind. Sometimes we find artifacts near nut trees, which they planted so they'd have food where they were going. If I could pin down one of these camps and protect it, excavate it, it would do a lot for the preservation of our history."

"There's a lot happening with the property right now," Evie said quietly. She shot a glance at Greg, remembering her mother's warning not to get anyone involved that would influence the outcome of the sale.

The archeologist nodded sadly. "I understand. Unfortunately, that's what everyone seems to say." He returned the arrowhead to Evie's palm.

"Thanks," Greg said and held the door for Evie to make an escape, the archeologist's eyes following her.

"I'm sorry...just with everything going on, I don't know how much I should look into this..." Evie ran a hand through her hair.

"I get it. You know what... I can look at your site when things settle down if you'd still like me to. I in no way want you to feel pressured. It sounds like your family has big decisions to make.”

Evie sighed. “Not us. Just my grandfather. And from a conversation I had with my mom on my way here, he’s already made his decision. I just...I don’t know if I can get excited about this only to lose it, you know?”

Greg smiled and pushed up his glasses. She couldn’t help but give him a halfhearted smile back.

“I understand, really I do. Can I buy you a coffee? You can tell me about this farm, and if you’re up for it, I could tell you a little about the gold mine I almost bought in Colorado before I found out it never produced gold.”

Evie laughed. “A gold mine without gold?”

“It’s as entertaining of a story as it sounds. There’s a great coffee shop in the next building if you don’t mind sitting with a bunch of college kids.”

One of said college students passed in the hall and shouted out a “Hi, Doctor Greg!” before carrying on. Greg waved back.

He shifted the heavy backpack on his shoulder as he waited for her to answer. “I promise I’ll still carry this to your car for you, too.”

Evie pursed her lips. “You think the coffee shop takes payment in gold?”

Greg laughed. Evie led the way down the stairs and out into the bright sunshine.

While they waited for their coffee, Evie's phone rang. Her throat tightened when she saw Pap's name. With a deep breath, she stepped away from Greg. Pap asked how things were going, and she filled him in on everything the geologist and his archeologist friend had said. It was hard not to relay their excitement. If Pap had made his decision, she didn't want to make him feel bad now. He listened patiently, then talked for a long moment more. Greg handed her coffee to her, his eyebrows raised in a quizzical expression. Evie could only look at him with wide eyes, her coffee cup frozen in mid-air.

Pap's voice echoed through the phone. "Evie, honey, you still there? What do you think?"

Chapter 34

Bill pulled his coat a little tighter around himself as he sat on his bench, the one that looked out over the entire farm from its little rise. The wind swirled around him gently, carrying leaves on its breath. Even though sunset was hours from now, he had the routine memorized. The light would turn golden. Then the colors would start splashing across the horizon, rimming the clouds in a showcase of heavenly brushstrokes. He didn't need a sunset, or a sunrise for that matter, to enjoy this view. The creek wove like a shining ribbon through the field, with the geese on the pond honking and splashing. The trees around the house were fully turned now, each standing like a bright golden flame. Even the barn, with its chipped paint, seemed nostalgic.

He watched a car pull in from the main road and wind towards the house, the gravel crunching. That looked to be Bess, coming to start prepping dinner. When she parked her car, she didn't look up the ridge towards him, but headed straight into the house as if she lived there.

She had for so many years. He hoped she would always feel that way.

A few minutes later, a truck pulled in, barreling down the gravel road far too quickly. Bill straightened a little. That had to be Jeth. Why had it taken him so long to call? Bill reached into his coat pocket and took out his flip phone. The screen was blank. Ugh, he'd forgotten to turn the blasted thing on again. He looked back up in time to watch Jeth jog into the house. A few minutes later he re-emerged, a hand shading his eyes as he scanned the fields. Finally, he looked up towards the rise, and Bill waved. Jeth immediately headed for him, his steps brisk.

He stopped a few feet away, bit his lip, and then silently sat next to Bill. Bill could feel the agitation rolling off his grandson, but wasn't sure how to proceed. If Jeth was going to be a man, he was going to have to broach the conversation like a man. After a minute or so, both men settled into the peace of the moment, listening to the whistle of wings as geese flew overhead.

Jeth plucked a leaf from the grass, its yellow veins spotted with flecks of red. He started folding it like origami. "Pap, the decision was yours. I should have come to you sooner. I should have been more involved here sooner."

Bill straightened and shifted his seat on the wooden bench. "If you had made an offer, what would you have said? What would have been fair to your sister and cousin?"

Jeth began ripping the leaf into tiny pieces. "I would have offered farm value, which I know is significantly less. But I don't want the bank to own the majority share. I don't want to risk losing it. I would have offered to split it three ways, amongst Evie, Marissa, and I. I would buy Marissa out so she could use the money to finish medical school and live her dream. Evie and I get along well enough; we could make it work as co-owners. She would make sure every inch of this place was operating to its potential. Then I'd quit my job—" He snorted and his words faded. "I actually did quit my job today. Clint made me choose between work and Sienna, so I asked her to marry me."

"You did? Did she say yes?"

"Of course." Jeth pulled the little box from his pocket and handed it to Bill.

Bill opened it and smiled at the rose-gold band with one of the shiniest diamonds he'd ever seen. "She's going to love it. Congratulations."

Jeth smiled, but it didn't quite reach his eyes. He was still too distracted. "Today's been a rollercoaster. I just wish it had all happened one day sooner." He shook his head. "Anyway, I also would have proposed that you stay in the farmhouse as long as possible. Free rent if you will, in exchange for taking the loss of profit. I'd be here more often to keep an eye on you, as would Sienna hopefully. And I already have my little house in town, so it's not like I'd need a place."

"How would you finance it?"

"Between what Sienna and I have saved, the bank loan would be manageable, particularly if we could get Evie on board as well. I'm going to keep engineering on my own to foot the major part of the bills." Jeth threw the pieces of the leaf like confetti to the ground. "None of that matters though, right? I'm too late. I'm sorry, Pap." His voice wavered, and he buried his head in his hands.

Bill felt his eyes tear up, and he wrapped an arm around Jeth. He had to clear his throat several times before he could speak.

"It's a good thing I didn't sign, then."

Jeth wiped his eyes, then slowly turned his head to look at Bill.

"What?"

"I didn't sign."

"But Pap, it's ten million dollars..."

"You going to take back everything you just said? Change your mind? Or you going to commit to this place like it's part of you? Like you deserve it as your inheritance?"

"I—yes. Yes, I want it."

"Then I'm gifting it to you and Evie. I want the two of you to buy out your sister. She's a responsible kid. She'll put her five hundred thousand to work in a beautiful way, I have no doubt."

"So Evie and I owe you…"

"You don't owe me. You each owe Marissa two-hundred fifty thousand. Then you just have to promise me you and Evie will get along. And I like the deal where I get to stay in my house. Very much." Bill smiled then, watching the shock roll in waves across Jeth's face. He gave his grandson a big squeeze. "If I were you, I'd let Evie claim the mountain. I just spoke with her, and she's on board with this whole thing. Turns out we do have a little gold, and she wants to start a tourist attraction rock-hounding and maybe even bring people in to study the area around the cabin as an archeological site. Sounds like she's met a geologist that is as excited as she is. There's something rare up there. She sounded more excited than I've heard in a long time."

"Evie, on the mountain?"

Bill laughed, the sound carrying across the farm. The geese added their song, honking down in the pond with earnest. "I think you two will figure out how to make it work."

Jeth threw his arms around Bill with a grip strong enough to make his ribs hurt. Maybe he'd make a farmer after all. Bill felt his grandson's shoulders shake and did his best to hold him. A minute later, Jeth pushed to his feet and let out a whoop. He turned to face the farm with a huge smile and spread his arms, his eyes closed. The wind blew, and Bill watched as the next generation claimed the land as his own.

Author's Note
Truth or Fiction

This is a story that simply burned to get out of my mind, and it is the fastest one I have written from concept to publication. After it was on paper and I could talk about it with friends, family, and readers, I realized just how timely this story is. Everyone driving down the highway, regardless of what state you live in, can witness the constant development, often at a huge scale. I'm a businesswoman that loves to see economic progress. That doesn't mean the development of farmland and natural spaces doesn't sadden me. Much like Pap's neighbors, I like the views out over those rolling cornfields. Couple that with exposure to the farmers of the families my sister and I married into, and I quickly could see just how complicated this entire dynamic between nature and progress is. I don't have the answers. It is my hope that this story makes you think. And as you think about the complexities of the situation, here is some clarification of things I discuss in the novel.

FARMING VERSUS DEVELOPMENT

The average American is at least three generations removed from the farm.

I was born even further disconnected from farming than that. Then I started a horse farm, which "real" farmers often don't count as farming. Then my sister and I both married farmers. Now I have seen firsthand how the rewarding, beautiful lifestyle contrasts to the sheer hard work and at times heartbreak that farming encompasses. The people in my life that farm, men and women alike, have a true passion for what they do. They take great pride in carrying on family traditions.

The joys, hardships, and economic decisions that this book presents are real. There are a lot of families, not just in Pennsylvania but around the country, facing tough questions. What is land worth? At what point is development a valuable economic benefit versus a threat to food security and a way of life? Would it be easier to sell out and get a job with benefits, shorter hours, and a higher wage? Every farmer and family is answering these questions in different ways. Many farms have shown creativity in order to increase productivity, diversify, and make the lifestyle profitable for future generations. It should be noted, however, that there is an alarming trend of small farms closing. Nationwide, according to the

USDA's 2022 report on Farms and Farm Land, we lost 9,350 farms in 2022 alone. Since 2015, we lost 61,190[1].

WAREHOUSES FOR LEASE

At the time of writing, a search of Realtor.com revealed twenty-seven warehouses for lease in Berks County, Pennsylvania, which is just south of Schuylkill. In a recent check, there are still twenty-six properties available for lease, ranging in size from 200 to 800,000 square feet. In Schuylkill County there are eleven for lease, the largest of which is a recently completed behemoth at 1,005,000 square feet. They are building more.

GOLD

Yes, gold has been found in Pennsylvania, usually in placer amounts. The Cornwall Mine in Lebanon County, an iron mine, found that they could process copper and gold as byproducts out of their ore. And yes, there are stories of nuggets being pulled out of some of the rivers,

---

1. "Farms and Land in Farms 2022 Summary." February 2023. https://downloads.usda.library.cornell.edu/usda-esmis/files/571 2m6524/bk129p580/2z10z22698/fnlo0223.pdf.

like the Susquehanna. That said, gold is still more rare in Pennsylvania than some of the western states. And through the story of the Gobbleberg treasure originates in the book *Old Schuylkill Tales* by Ella Zerbey Elliot, to date, no one knows what mountain the old Indian was referring to. Thus, no, no giant boulders containing gold nuggets have been found. Yet.

PENNSYLVANIA NATIVE AMERICANS

Prior to 1749, treaties between settlers and tribes drew the line along the Blue Mountain, which today marks the boundary between Lehigh, Berks, and Schuylkill Counties. Another land purchase was made in 1752, allowing the settlers to expand over the mountain into what is now Schuylkill County. The tribes protested. The 1750s became years of conflict, necessitating the building of a series of forts along the base of the mountain, today marked with historical signs. The story of the Alspachs is true (and I encourage you to read the full story in the *Blue Book of Schuylkill County*), and though the Fogels in my story are purely a product of my imagination, there are a few families still in the area that can trace their roots back to that time. Tensions quieted by the 1780s, when treaty lines were yet again redrawn even further to the west. The Pennsylvania tribes were displaced or absorbed into the European settlements. There are some sites working to preserve the long history of Native Americans here in Pennsylvania, though as more and more areas of bare land are developed, one has to wonder how much history we are losing.

THE SCHUYLKILL COUNTY DIALECT

I kept the dialog in this novel in our local Schuylkill County dialect, which is a bit unique. Even as recently as my great-grandparents, locals spoke Pennsylvania Dutch. My husband's grandfather knows enough of it to tease you, and we have nearby Amish communities that still speak it. If you haven't heard of Pennsylvania Dutch, I encourage you to look into it. It is a unique language based on the German (Deutsch) of our ancestors. Laced into the old "dutchy" grammar that destroys conventional use of prepositions, we have an influence of Appalachian dialects, tied into our coal mining roots. Nothin', kinda, and yous all stem from this. This can be read as a bit of a southern influence, but it's not quite that either. Aren't dialects fascinating?

I hope you enjoyed *The Last Farm*. Share your own stories of family, farming, or the development boom at www.authorjastein.com.

Acknowledgements

I had an amazing team behind this book, and I am forever grateful to each and every one of them.

For one, my husband, my very own farmer/engineer, who not only watched the TV on mute the nights when I was deep into the story but answered dozens of questions. You can thank him for telling me what crops are harvested in fall, which tractor part would need to be fabricated in a machine shop, and what kind of engineer builds warehouses. Thank you, honey!

Then there was my awesome team of critique partners. Richard, Emily M., Patricia, Lavada, Emily H., Chad, and Mom— thank you for putting up with all my typos to see the story underneath and for all your help making it shine.

Thank you to my wonderful editor Gail Delaney, who somehow manages to find all my silly grammatical mistakes while remaining positive about my work.

A big thanks to the ladies at the Orwigsburg Library, who volunteered to be my advance readers and remain some of my biggest cheerleaders as I branch into a new genre. You are the best.

To all the readers of *Knightess*, thank you for encouraging me to keep writing! I hope you enjoy this story, too.

And finally, thank you to my two farmer brother-in-laws and all the other farmers out there. Your passion for what you do is inspiring, and may we recognize your hard work and be grateful for the food you put on our tables.

About the author

J.A. Stein grew up exploring the woods near her Pennsylvania home. It was there that she learned to let her imagination run wild. Love of the outdoors led to a career with horses, now shared with her passion for writing. Her debut novel *Knightess* is the winner of a 2023 Royal Dragonfly Book Award. Stein runs a horse farm just a few miles from her husband's family farm, where they are still running seven generations strong.

Reviews of this book are greatly appreciated and can be left on any retailer site, as well as BookBub and Goodreads. For bonus chapters and upcoming events please subscribe at www.authorjastein.com. Looking for social media? You won't find it. Stein believes time is better spent reading a book than a news feed. Then there is more time left over to experience your own story. That said, we'll pretend Youtube isn't social media. For glimpses of her farm, book trailers, and inspiration, check out her channel @AuthorJAStein.

www.ingramcontent.com/pod-product-compliance
Lightning Source LLC
Chambersburg PA
CBHW021140310726
48971CB00002B/421